THE PROFESSIONAL ENGINEERING CAREER DEVELOPMENT SERIES

CONSULTING EDITORS

Dr. John J. McKetta, Jr., Executive Vice Chancellor for Academic Affairs, The University of Texas System

Dr. Maurits Dekker

EDITORIAL ADVISORY COMMITTEE

Dr. Walter O. Carlson, Acting Dean of Engineering College, Georgia Institute of Technology

Mr. R. E. Carroll, Director, Continuing Engineering Education, The University of Michigan

Mr. William W. Ellis, Director, Post College Professional Education, Carnegie-Mellon University

Dr. Gerald L. Esterson, Director, Division of Continuing Professional Education, Washington University

Dr. L. Dale Harris, Associate Dean, College of Engineering, The University of Utah

Dr. James E. Holte, Director, Continuing Education in Engineering and Science, University of Minnesota

Dr. Russell R. O'Neill, Associate Dean and Professor, School of Engineering and Applied Science, University of California at Los Angeles

ENGINEERING PROFESSION ADVISORY GROUP

OCEANOGRAPHY FOR PRACTICING ENGINEERS

Luis R. A. Capurro

**Department of Oceanography,
College of Geosciences, Texas A & M University**

BARNES & NOBLE, INC. NEW YORK

Publishers • Booksellers • Since 1873

Distributed

In Canada
by the Ryerson Press, Toronto

In Australia and New Zealand
by Hicks, Smith & Sons Pty. Ltd.,
Sydney and Wellington

*In the United Kingdom, Europe,
and South Africa*
by Chapman & Hall Ltd., London

Printed in the United States of America

Preface

A look back to the last two decades indicates that the world is being subjected to unprecedented social and technological revolutions. Too much is happening too fast in both areas.

Although it is clear that the solutions to the rising social problems will come from science, it is curious that each one of the answers is in some way connected to the sea. The image of the sea has emerged as a potent force which spreads across the entire spectrum of future human affairs.

The implication of the sea in the rapidly developing world society is clearly seen in a new concept that is becoming the main motive of many oceanographic programs—*the effective use of the sea* by man. This involves the use of the ocean for all purposes to which man now puts the terrestrial environment: commerce, industry, recreation, and settlement, as well as knowledge and understanding.

Thus man is beginning the conquest of a new environment. The technology is under accelerated development. The task is formidable, but the final results will be very rewarding.

Acknowledgments

I want to express my sincere appreciation to William J. Merrell, Jr. for his critical review of the manuscript and for his constructive criticism. A considerable amount of time and effort has been expended in the preparation of the book by other persons. I would like to thank Hector R. Cornelio for the arduous task of preparing the manuscript for publication, to June Hagler for typing each of the several drafts of the manuscript, and to Mrs. Deana Rupley for her secretarial assistance.

Contents

1
Introduction

A look at the present effort in marine science activities shows a new pattern that reflects how society is reacting to meet its expanding needs. Emphasis is changing from what was largely a scientific study of the sea to its more effective use. The agent for the critical transfer of scientific results to practical application is ocean engineering, which is emerging as a recognizable field of endeavor in most maritime countries of the world. It is not an easy task to determine the boundary between marine science and ocean engineering. A characteristic of the environmental sciences is that many of their scientific programs are so extensive or have such a degree of sophistication that the engineering requirements are necessary for their successful implementation. Therefore, an expanded spectrum of ocean technologies is a dominant requirement not only for the effective occupation of the sea but also for providing an adequate infrastructure for the scientific study of the environment.

The marine environment presents two challenging problems for the engineer: (1) design and development of the tools and instruments to observe and measure ocean phenomena and (2) transfer of the scientific discoveries into practical use of the oceans. The objective of engineering research is to obtain data and develop the theoretical foundations for the application of basic engineering principles into the oceanic environment.

The acquisition of data for the observation and measurement of ocean phenomena is not an easy task if one considers the nature and limitations of the marine environment.

The medium to be observed is an almost impenetrable curtain to electromagnetic radiation. Visible light penetrates the sea surface only a short distance, after which it is scattered and attenuated in such a degree that images cannot be recognized.

It is, however, more permeable to acoustic energy, which can be propagated for considerable distances, although the normal density stratification of the sea bends and distorts the sound waves, making the ocean behave like a poorly designed auditorium. The marine environment is in continuous motion, and its spectrum of variations is not well known. This condition of the ocean creates mechanical stresses on undersea structures, platforms, and vehicles quite different from those found on land. Also, the variations are of such magnitude that what really count are the maximum forces that structures might have to withstand for very short periods from currents, surface or internal wave motion, or seismic shocks. Moreover, the ocean is a highly corrosive environment. Electrolytic and biological activity create problems quite different from those found in the air or on land. The living organisms attach to submerged structures, make noise, bore in them, excavate under them, and generate a series of unpleasant problems.

This complex and changing environment has been observed and measured chiefly with remote self-operating instruments used from surface ships, buoys, or fixed platforms. Lately, the development of the self-contained underwater breathing apparatus and free-floating underwater balloons has enabled scientists to make short visits to the upper layers of the sea and to abyssal depths. But the scientific information thus obtained is far from satisfactory.

In recent years, the techniques and instruments used to observe the ocean have been rapidly improved, thanks to the great advances in electronics and related fields of engineering. The rate at which some types of data can be collected has reached the point that they can be studied only after being combined and correlated in large computing machines. However, the need for new devices and specialized instruments for effective ocean exploration and research is still a demanding requirement if one wishes to comprehend the marine environment.

The other responsibility of ocean engineering is to provide the means to convert the scientific information on the marine phenomena on useful applications for the benefit of man. It is not presumptuous to say that ocean engineering is a key to the development of the ocean's resources, to maritime transportation, to defense, and to enhancement of the use of shores and harbors to meet society's expanding needs.

Evidently, successful ocean engineering will evolve from classic engineering principles, modified to take into account the different ambient factors and forces of the ocean and oriented to the activities man wants to accomplish in the sea. The opportunities in marine engineering will be concerned with propulsion, materials, sources of energy, structures, navigation and positioning, communications, identification of objects, buoys, and so forth. Engineering design must take into consideration the classic parameters such as cost, safety, reliability, availability of components, and ease of maintenance; but implicit, too, is the systems approach to problem solving, where the following factors, typical of the marine environment, will be taken into account: sea surface motion, tides, and currents; wave impact and wind loadings; heavy hydrostatic pressures; large buoyant forces; opacity of sea water to electromagnetic energy; high attenuation and scattering of light energy; high conductivity of sounds; lack of gaseous oxygen for man or chemical combustion; presence of all common elements in sea water; variable two-phase nature of the water–bottom interface; and severe corrosion and fouling.

Considerable ocean engineering experience has been gained during the last decade in highly industrialized countries, mostly from civilian economy and defense requirements. The activities of the gas and oil industry have extended to greater depths and into a broader range of geographic and climatic conditions. This, as a result of the heavy storm losses in different regions of the world, has accentuated intrinsic problems related to the installation and maintenance of offshore towers. Consideration of underwater mining in coastal waters, on the continental shelf, and even on the deep sea floor has promoted research in new engineering problems. Ships and submarines, undersea cables and tunnels, and coastal protection are all examples of successful ocean engineering. This experience has been obtained mostly on an empirical basis, and although technical developments based on trial-and-error experience may continue to move ahead of the theoretical understanding in many cases, this approach has resulted in a fragmentary view of the problems that should be organized to make better use of the oceans.

There is a lack of literature dealing with ocean engineering problems. There are insufficient engineering data, and there is not enough broad and generalized familiarity with an insight into the ocean; and, therefore, there is a lack of professional preparation to deal with the special engineering problems imposed by the marine environment.

The time has come to establish an identifiable general field of ocean engineering that will have the characteristics of other fields of engineering. The increasing numbers of universities with curricula in marine engineering, the journals dealing with marine technology, and the more frequent special meetings on ocean engineering are a clear indication that the time is ripe to begin such an undertaking.

The purpose of this book is to try to convey to the practicing engineer some of the characteristics of the marine environment that have a bearing on his envisaged professional activities if the effective use of the sea becomes a reality. The strategy followed in the presentation of this work is to overemphasize, when necessary, the particular characteristics of a large-scale natural system such as the ocean and atmospheric environments, to compare phenomena in both fluid media when convenient for clarification, and to comment in detail on the application of certain well-known properties, equations, and principles of classical physics on a geophysical scale.

Most of the environmental factors and phenomena that play important roles in engineering research apply to the watery environment. Moreover, fluid mechanics covers a broad spectrum of the engineer's interests. It arises in every branch of engineering: waterworks, engines and power plants and aircraft, chemical plants, building, transportation, rockets, and many other areas of activity.

Because of this considerable interest in fluid dynamics and because some elaboration is required to explain the differences between fluid dynamics and *geophysical fluid dynamics* (fluid phenomena applied to a natural large-scale system) Chapters 3, 4, and 5 have been treated in considerable detail.

2

The Marine Environment

CHARACTERISTICS OF THE WORLD OCEAN

The oceans of our planet constitute a natural large-scale system of astonishing complexity. They are one of the three great environments of the earth which continuously interact with each other through their natural boundaries, the sea floor and the air–sea interface. The degree of complexity can best be illustrated if one considers that in the physical sense the environment to be studied is characterized by "a complex interdependent fluid phenomena produced by the combined action of nonuniform distribution of sunlight over the earth (known as short-wave radiation), the equally energetic but more uniform radiation of earth heat into space (long-wave radiation), together with rotational and gravitational forces." Someone expressed in a simpler form the complexity of both fluid environments (air and water). They are "turbulent, differentially heated fluids in restless motion, upon a rotating spheroidal planet." Therefore, in dealing with the specific oceans of the earth, it is important to keep in mind that they are unideal, ungeometric, and inhomogeneous. To describe them and to understand their behavior, a particular strategy common to most natural planetary systems must be followed: to subject the specific object of study—the sea—to a high degree of empiricism, observation, and qualitative thinking, and to assemble the analytical results of disciplinary science into a coherent understanding.

The specific subject of oceanography is the *world ocean*. It seems fitting, before getting involved in the discussion of the scope of the science, the different fields of specialization, and how the oceanographers organize their research, to meet the challenge of understanding the oceans, to amplify the characteristics of the large-scale natural system known as the world ocean.

The water of the oceans, together with the air of the atmosphere and the rock of the solid earth, make up the planet Earth. These three environments are also known as *hydrosphere, atmosphere,* and *lithosphere,* respectively. The most logical arrangement would be to have those environments in concentric shells in decreasing order of density as the ratio of the shells increase. However, this is not the case, as the solid earth is very uneven, and depressions and elevations disturb its regular shape. The lithosphere has internal energy at work, and powerful forces within the earth's mantle have elevated the continents which constrained the waters into large depressions that constitute the individual oceans. The only oceans with zonal continuity are the southern ocean around Antarctica and the Arctic Ocean in the north. Some dynamic process in the mantle, under the ocean, maintains this kind of arrangement.

The transition from one environment to another does not take place in a gradual way; it is rather abrupt. The transition zone between the solid earth and the water is the sea bottom. The density changes from around 2.5 to 1.06 g per cu cm. The upper boundary between the water and the air is the sea surface, and the density varies from 1.3 to 0.0013 g per cu cm. These two interfaces are of great importance to oceanography. The waters of the oceans lie as a single continuous mass within these two boundaries; all the energy gained or lost from the oceans must pass through these boundaries, and they are responsible for the phenomena taking place in the water mass.

The earth's surface is mostly oceanic. The ratio of land to sea is 1:2.43, or 29.20:70.80 percent. The arrangements of the continental land masses outline the irregular distribution of the sea. The three major water bodies, the Atlantic, Indian, and Pacific oceans, are connected, forming a continuous ocean belt known as the southern ocean or the Antarctic Ocean. The impressive size of the world ocean and the continuity of the waters give to the marine environment the characteristic of being a large-scale natural system, and to oceanography the

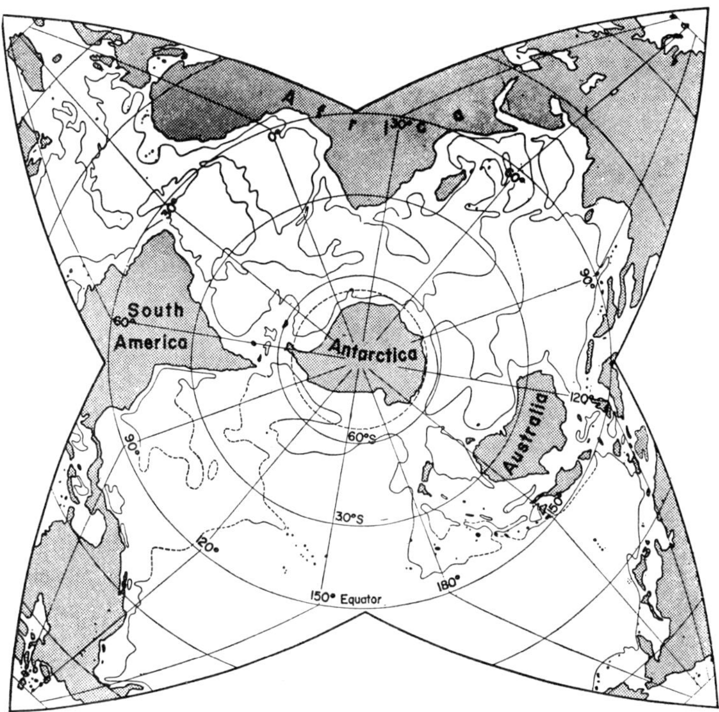

Fig. 2.1   Steinhauer star projection showing the distribution of oceans and continents. (From A. Defant, *Physical Oceanography*, Vol. I, New York: Pergamon Press, 1961.)

scope of a planetary science. Figure 2.1 shows the relative size and degree of connection of the oceans.

The boundaries of the ocean waters are of great importance if one considers that all energy has to pass through them. The boundary with a higher degree of interaction and through which most of the driving forces that generate ocean phenomena are transmitted is the *ocean–air interface* or the *sea level.* In a physical sense, it is a free boundary that may assume different forms at different times under the influence of the external or internal forces. If the atmosphere were removed and no gravitational effects from the moon and sun were present, the sea level would be subjected only to the action of the earth's gravitational pull and the centrifugal force of the earth's rotation. For a homogeneous land, this *ideal sea level* will coincide with the surface of a rotational ellipsoid. However, the solid earth is not homogeneous and the irregularities of the gravitational field due to the irregular mass distribution of the earth's crust create irregularities on the surface of the rotational ellipsoid, and this new surface is called the *geoid*. The topography of the geoid in relation to the rotational ellipsoid is such that it is higher over the continents and lower over the oceans. The magnitude of these deviations does not exceed ± 100 m. If the atmospheric and attractive forces are considered, the geoid sea level will begin to oscillate at different frequencies around a mean value known as *mean sea level*. The power spectrum of these oscillations is shown and discussed in Chapter 5. The growing interest in long-range weather forecasting has considerably increased the research activity on the nature and behavior of this interface.

The other boundary layer, commonly known as *sea bottom* or *benthic boundary*, is less variable than the surface one. The detailed nature of this boundary and whether its characteristics result primarily from physical or biological processes are not well known. The topography of the sea bottom is far from uniform, and features similar to the ones observed on land are present on the sea floor. Although this boundary is less variable than the sea surface, important dynamic processes take place in the shallow water province, particularly in the nearshore region. Because the crust is very thin beneath the deep sea floor, the benthic boundary is now the base for studying the earth below.

Another important characteristic of the marine environment is the high degree of variability of the physical and biological processes taking place there, especially in the upper layers of the sea. The oceans are restless—every drop of sea water in the world is constantly in motion. The spectrum of variations in the motions in the sea is a broad one. There is a great variety of motions, from the familiar surface waves to the slow currents deep within the sea, from the majestic flow of oceanic currents such as the Gulf Stream in the western North Atlantic and the Kuroshio or Black Current off Japan that carry impressive amounts of water, to the swift tidal streams of harbor mouths. Much of the motion is turbulent, but in all cases the water particles follow nearly horizontal paths. This is so because, on a planetary scale, the oceans are only a thin stratified film over the globe.

Time-series current measurements taken from moored buoys have shed considerable light on the time distribution of horizontal motion but little on spatial variations. Although the scale of horizontal particle displacement is known, the coherence over larger distances has not been determined. Only on the continental shelf and in constricted regions of the ocean are tidal currents found to be strong in comparison with inertial motions.

For variations of a period below 10 hr (frequency, 0.1 cycles per hr), the energy density of velocity fluctuations decreases with frequency, and the corresponding horizontal spatial scales are 10 km and less, and vertical scales are not known. For periods greater than 24 hr (below 1 cycle per day), the energy tends to increase with decreasing frequency, but very little is known about the temporal and spatial distributions. The vertical motions of water, although of much smaller size than the horizontal ones, are of great importance to many marine phenomena. The transfer of cold and dense nutrient-rich water from intermediate ocean depths to the upper wind-stirred layer of the ocean is not only of great importance in the biological productivity of the sea, but it also has profound effects on weather and climate because of the changes introduced to the sea surface temperature field by the cold upwelled waters. Because of the low value of the upward motion, it has not been possible to make direct measurements of vertical velocities but only to infer their rates from changes in the distribution of properties.

Matter and energy are also transported by nonconvective processes, such as turbulence, which results in eddy diffusion and mixing. These processes are present in many different scales and are generated by different causes. In the upper layers of the sea, the winds and waves are the main energetic sources for turbulence, but the manner in which the energy is transferred and the energy budget are not well known. In the oceanic stratosphere, below the pycnocline, strong irregular currents are found, but their interactions with the steady-current systems are not known. It has not been determined whether turbulence is a result of the mean motion or whether the mean motion results from bracketing turbulent energy. It is possible to show that tidal energy is mostly dissipated through turbulence in the shallow seas, yet a considerable fraction provides the energy for maintaining internal waves in the deep sea. It has been possible to measure turbulent velocity fluctuations of periods between 0.1 to 100 sec (0.01 to 10 cycles per sec) in a sampling volume having dimensions of the order of 1 cu cm and transient eddies of sizes up to 50–100 km. All of these turbulent motions must be better known before a sensible theory of ocean circulation can be developed.

The oceans, like the air, exhibit the unique characteristic of being stratified, that is, divided into layers separated by density contrasts. This property is of fundamental importance in most oceanic processes and complicates considerably their theoretical formulation. This is, in a way, a contradiction to what might be expected on the basis of the continual motion of the waters, where mixing would create uniformity in property distribution. In fact, this is not the case. Other processes bring about differences between water masses, and these differences are maintained by the stratification of the sea. If water near the bottom is sampled at the equatorial Atlantic, it will be found that the temperature is close to freezing. This very dense water sunk from the surface near Antarctica thousands of years ago and retained its antarctic temperature as it slowly moved northward. As it moves and spreads, it mixes and gets heated and eventually gets to the surface in the North Atlantic. The deep and intermediate circulations are responsible for the important characteristic of *stability* in which work has to be done against gravity to bring water from the depths to the surface layers. Stability is the property that controls vertical mixing, circulation, organic production, and other oceanic processes.

The oceans provide a gentle physical environment for living things and there dwell hundreds of thousands of different kinds of marine organisms, ranging from the microscopic algae—the phytoplankton, which are the basic food source for all marine animals—to the largest animals that ever lived, the giant whales.

The complete aggregation of organisms of all kinds, together with their activities, constitute what is known as the *organic environment*. Being a part of the whole series of factors that control the existence of the individuals, they form the vital portion of the environment.

Biologists have been fascinated by the wonderful complexity of the ways of life in the sea. One of the classical fields of biology is marine biology, which deals with the physiology, life histories, distribution, behavior, and evolution of marine organisms. This area of research, together with biological oceanography, is becoming more and more important because of the rapid expansion of fisheries throughout the oceans of the world.

Ecological relations are well exemplified in different regions of the world ocean where differences in the properties of the waters are reflected in variations of fertility. A most dramatic example of highly productive waters is the narrow strip of ocean 100 miles wide and 1000 miles long off Peru in the South Pacific. The water near the surface forms a green permanent pasture that sustains one of the greatest fishing grounds in the world.

The marine environment is less efficient than the land environment as an organic producer because of the elaborate food chain in the ocean that reduces considerably the final yield in relation to the great primary production available. A quick look at the food web in the marine environment shows the reasons for the reduced final yield and suggests some steps to improve the situation. On land, the food energy available to man is likely far greater than at sea because the primary product, the land plants, are more available. In the sea, the primary producer is the phytoplankton—microscopic floating plants—which is passed in a number of steps through the food chain to increasingly larger organisms. Each of these steps or trophic transformation yields about 10–15 percent of the respective inputs. Therefore, the fish which are well ahead in the food chain will be available as a small portion of the initial productivity available. The higher-level predators, such as tuna, cod, shark, and squid, represent only a small part of the available protein. It is clear that a great step forward to increase the organic yield of the sea for man's use is not to produce more plants but to eliminate steps in the food chain.

THE SCIENCE OF OCEANOGRAPHY

The specific subject of the oceanographic science, the sea, has been briefly described. Some of the important characteristics and challenging problems of the marine environment have been presented. It is now convenient to see what tools are available and how the oceanographers deal with the task of understanding the oceans.

Oceanography is defined in various ways depending on the concern of the definer. For the scientist, oceanography is the scientific study of that part of the earth which is covered with sea water. His aim is to understand all aspects of the oceans: the properties and behavior of the ocean waters; the nature of living organisms in the marine environment; the interactions between the waters and their boundaries, the air above them and the solid earth beneath; the shape and structure of the ocean basins; the economic and technical potentialities of the oceans and their roles as a part of the earth's outer covering. Where sea water touches the solid crust, oceanography enters the domain of geology; where it reflects sunlight, evaporates in the atmosphere, or exerts a drag on the winds, oceanography merges with meteorology; where marine forms of life exist, oceanography is joined with biology; and where man must combat or find uses for the sea or sea water itself, oceanography gets in touch with engineering and technology. It is a basically interdisciplinary and relatively unspecialized field concerned with the sea as a major part of the human environment.

Oceanography is not a universal science. The implications of this statement are that, in principle, it can be said that the laws of physics hold good everywhere and at all times, and the chemist's discoveries are valid whenever and wherever molecules can exist; but oceanography consists of the application of physics, chemistry, biology, and mathematics to the study of a particular object in space and time—the oceans of the planet. It is a typical environmental science which shares with other environmental sciences the peculiarity that the present condition of its object of study has been largely determined by events in the distant past.

As a planetary science, it also shares with other planetary sciences the difficulties of reproducing natural conditions in laboratory models. The physicist can make his own world by conducting controlled experiments in his laboratory. The applied mathematician can analyze the behavior of imaginary fluids, but the oceanographer on board his ship must look at the sea as it actually exists. For the physicist and the applied mathematician, there are many possible oceans. For the oceanographer, there is only one, the earth ocean, with all its long and difficult history.

The statement that oceanography is a planetary science does not refer only to its planetary size. Also involved is the concept of the continuity of the marine environment. Like the air in the atmosphere, the ocean waters are indivisible, and events in any part of the sea eventually have profound effects at great distances. The Atlantic *deep water* is formed during winter in the North Atlantic. Here, high-salinity waters, which are cooled at the surface, attain a density which is sufficiently high to sink and flood the depths of the North Atlantic Basin and flow southward overriding the more dense Atlantic *bottom water*, which was also formed at the surface near the Antarctic continent in the Weddell Sea. After moving through the South Atlantic, it climbs to shallower depths in the southern ocean at the *polar oceanic front (antarctic convergence)*, where it becomes the source of the impressive water mass known as *antarctic circumpolar water*, which flows all around Antarctica. Moreover, water which is formed at the surface in antarctic regions will show up in the northern parts of the oceans.

For the oceanographers, the classical division of the oceans on the basis of the continental barriers that partially enclose them is of limited value. The ocean, thus defined, has a more geographic meaning. Large-scale oceanographic phenomena must be studied considering the world's oceans as a whole. A similar conception of a global atmospheric physical system is applied by the meteorologists to study large-scale weather phenomena.

The classical subdivision of the science of oceanography is in the three main branches: physical oceanography, chemical oceanography, and biological oceanography. However, this subdivision is restrictive. Today, it is adequate to include marine meteorology, submarine geology and geophysics, marine geodesy, and certain branches of engineering as fundamental parts of the science.

Physical oceanography is concerned with the physical properties of the ocean waters and with the transport of energy, momentum, and matter. The major activities of the physical oceanographer consist of (1) direct observation of the oceans and preparation of synoptic charts of oceanographic elements and properties, and (2) theoretical study of the physical processes that might be expected to explain the observed behavior of the oceans. The first aspect is the typical task of the descriptive oceanographer; and, as such, it is a branch of physical geography. The second is a branch of theoretical physics. Neither can stand without chemical and biological information as a part of the description of the oceans and as a cross-check on the validity of the physical reasoning.

The role of the chemical oceanographer is nearly central, because his work is as indispensable to the physical as it is to the biological studies. Through chemical analysis, it has been learned that the composition of the salt dissolved in sea water is, in general terms, nearly the same regardless of its total concentration. This implies that the waters of the oceans are well mixed, a process that must be sustained by the presence of various classes of motions. Chemistry also contributes to the indirect determination of the density of sea water with the precision required by the physical oceanographers for the computation of the field of motion from the field of density. Near the coast, especially where rivers flow into the ocean, the *constancy of composition* of sea water cannot be assumed to hold even approximately. Dissolved inorganic and organic materials from rivers, industrial wastes, and ground water constitute an important environmental influence on the productivity of organisms in the sea. These problems make the work of the chemist indistinguishable from that of the biologist.

Two aspects of the study of biological things and processes of the sea need to be distinguished: marine biology and biological oceanography. Marine biology deals with the understanding of biological principles and processes that apply to living things; the ocean environment is largely incidental and various organisms or processes are chosen because they can fit the particular line of research well. Biological oceanography, on the contrary, is concerned with the biological environment, that is, with marine organisms, as part of the total oceanic system, and with the ocean as a habitat for life. It aims to understand the interactions of organisms with their environment and with each other. Although both disciplines have much in common, they have important differences in purposes, scope, and methodology. Biologists, especially the ecologists, often work in association with both chemists and physicists to determine the relations between organisms and their environment. Biologists are interested in determining the organic and inorganic intake of organisms, the cycles by which these materials are returned to supply future generations, and the trajectories of these materials as they go through the food cycle of the sea.

Marine meteorology or meteorological oceanography has received considerable attention during the last decade, and it is becoming an important branch in the sciences of oceanography and meteorology. The possibility of predicting large-scale atmospheric behavior for periods longer than a day or two depends on successful measurements of vertical fluxes of heat, momentum, and water vapor both on land and sea. The spectral structure of atmospheric turbulence is being determined, and direct measurements of vertical fluxes are being made with rapidly responding sensors mounted on fixed platforms, aircraft, or submarines. The behavior of the air–sea boundary constitutes the central problem in meteorological oceanography.

With the increasing emphasis on the effective use of the sea, which implies the observation of the marine environment and the exploitation of its potentials, it is difficult to define the scope of the branch of ocean engineering. Evidently, engineering is not an end, but a means for tools, techniques, facilities, and services that must match requirements for research and for applications. The main problem in ocean engineering is to apply the classic engineering principles and accordingly modify them to the different environmental factors and forces of the ocean and to develop the means for carrying on the tasks man wants to accomplish in the sea. The environmental factors that are being taken into account in the systems approach to problem solving are sea surface motion, wave impact, and wind loadings; heavy hydrostatic pressures, large buoyant forces, opacity of sea water to electromagnetic energy, high attenuation and scattering of light energy, high conductivity of sounds, lack of gaseous oxygen for man or chemical combustion, pres-

ence of all common elements in sea water, variable two-phase
nature of water–bottom interface, severe corrosion and fouling, and others of lesser importance. These typical characteristics of the marine environment create a real challenge to
classic engineering. Ocean engineering, a relative newcomer
in the technical world, must meet this extra challenge if it is
to mature concurrently with expanding marine requirements.

SUMMARY

The marine environment constitutes a large-scale natural system, where sea water—one of the geophysical fluids—plays
the dominant role. It is commonly referred to as the world
ocean and is characterized by its impressive size, the continuity
of its medium, its permanent interaction with the other two
geophysical media—the atmosphere and the lithosphere—and
the high degree of variability of many of the phenomena taking
place in it.

Oceanography is the science responsible for understanding
the ocean's behavior. Practically all disciplines may get involved to a lesser or greater degree in some particular aspect of
the marine environment. The description of the physical aspect
of the world ocean requires the participation of classic disciplines such as mathematics, physics, chemistry, geology,
meteorology, and even biology. However, because of the real
nature of the marine environment—a large-scale natural system—the application of those sciences has to be made on a
geophysical scale; and it is this aspect that requires the mathematician, physicist, or chemist to become acquainted with the
tools used to work in such a large scale.

The effective use of the ocean by man is largely an engineering enterprise. The necessary tool to put the scientific knowledge of the marine environment into practical use is the new
field of endeavor, ocean engineering. The conquest of this new
frontier requires an understanding of the environment which
generally is provided by the oceanographer. However, the
oceanographer needs the cooperation of the engineer to
understand the world ocean. The simple observation of such
a huge and variable environment puts such a heavy demand on
the scientist that he is unable to cope with it. The role of the

oceanographic engineer is to participate in the design of the system that will describe the ocean phenomena by designing the proper observation platforms, techniques, and instrumentation, and to store and process the impressive amount of data to be collected. This great contribution from the engineer will help in the solution of the basic problem—to comprehend the behavior of the world ocean. When this is accomplished, they will be able to go into the next phase—the effective use of the ocean by men.

3
Characteristics of Sea Water

The elements that constitute the physical marine environment are the sea water itself and its natural solid boundaries, the ocean floors. The description of such an inorganic environment comprises the knowledge of the physical properties and chemical constituents of the sea water, their distribution and concentration in space and time, the movement of the water, and the nature of the ocean floors.

The first striking fact about the liquid element is that the properties of pure water are unique in comparison with other liquids, and the nature of our natural physical environment is conditioned by the anomalous characteristics of the water. The presence of salts do not considerably change this unique behavior of water. The high values of heat capacity, latent heat of fusion and evaporation, surface tension, heat conduction, dissolving power, dielectric constant, and transparency have considerable bearing in conditioning the physical-biological environment. The explanation of some of these anomalies has to be found in the structure of the water molecule itself, the degree of polymerization and the presence of heavy isotopes of hydrogen and oxygen.

The presence of dissolved salts makes the sea water a solution and its behavior can be studied, making use of the body of knowledge in the field of solutions. As such, it has some unique properties that also affect the natural environment, namely vapor-pressure lowering, freezing-point depression, boiling-point elevation, and osmotic pressure.

However, oceanography deals with sea water as it occurs in nature, filling the basins of the world ocean and interacting with the other planetary media through the boundary layers. The presence of minute suspended particles and the different scales of motion affect several important processes. The absorption of light is considerably increased owing to the scattering produced by the suspended matter. The processes of heat conduction, chemical diffusion, and transfer of momentum are considerably altered because of the turbulent characteristics of the water motion.

It is common to refer to our planet as a water planet. One of the questions not yet well answered is the one related to the origin of the vast quantity of water present in the world ocean (about 1 billion cu km) and of the salt concentration (about 3 percent). The information available, particularly the studies of volcanic gases, shows that both the water of the oceans and the halogens might have been exsolved from the rocks forming the earth's crust. As for the sodium and other metallic ions, they can be accounted for by the chemical weathering of igneous rocks by precipitated water which then carries its dissolved mineral load to the sea.

In addition to the physical and chemical properties of sea water, there are other characteristics that result from the magnitude of the ocean itself, its great depth and its expanse. Taken in its entirety as an environment, we are at first impressed by the wide range in the value of the oceanographic elements and the concentration of some properties.

In the open ocean, the temperature ranges from about $-2°C$ to $+30°C$. The lower limit is determined by the formation of ice and the upper limit is conditioned by processes of radiation and exchange of heat with the atmosphere. The salinity in oceanic regions is generally between 33 and 37 ppm. At the surface, where most of the changes take place, it is affected by precipitation and evaporation. Near the mouths of the rivers it may be considerably less; and in places of high evaporation, like the Caspian or Red sea, salinities can exceed 40 ppm. The pressure in the oceans is essentially a function of depth, and its range will be from zero at the surface to over 1000 kg per sq cm at the bottom of the ocean basins. As the numerical value in decibars nearly equals the depth in meters, the range will vary from zero to over 10,000 decibars.

The mean values of the temperature, salinity, and density in the world ocean are $3.52°C$, 34.72 ppm, and 1.0276 g per cu cm, respectively. The precise determination of the spatial distributions of the oceanographic elements is a major achievement of modern oceanography and its observational technique.

Much of the knowledge and understanding of the ocean's behavior is obtained through the application of the conservation laws of mass, momentum, and energy to processes occurring in the ocean. To formulate these laws, it is necessary to know certain properties of the fluid medium, such as density, specific heat, viscosity, and other physical and chemical characteristics, as functions of temperature, pressure, and salt content.

The incentive for much of the early studies of sea water stemmed from the need to be able to apply the above laws with precision in terms of observations taken in the field. This was particularly true in the case of momentum equations with the development of methods of computing the field of motion from the field of density in the ocean. Because of this motivation, physical properties entering into these equations have been studied more intensely than others. A number of properties have been studied because of their importance to life in the sea. Others were investigated for technological reasons and for the development of special techniques for the analysis of sea water.

SALINITY

The ocean basins are filled with sea water, and what characterizes sea water is its salinity. The concept of salinity aims to give the value of the dissolved material per kilogram of sea water and has been defined in different ways, according to the techniques used to determine it. The classic definition of salinity is "the total amount of solid material in grams in 1 kilogram of sea water when all the carbonate has been converted to oxide, the bromine and iodine replaced by chlorine, and all organic material completely oxidized." This definition reflects the original procedure used to determine the amount of dissolved solids in sea water, but in practice this procedure is difficult to carry out with high precision, and an empirical relation between salinity and chlorinity has been used as a working definition:

$$\text{Salinity (ppm)} = 0.030 + 1.8050 \text{ chlorinity (ppm)}$$

This relation, which is another way of defining salinity, has been very useful because of the relative constancy of proportions of the major constituents of sea water, and because of the availability of a precise chemical method for determining chlorinity. However, this relation has its limitations because it is based on few salinity determinations and also because the constant 0.030 results from the use of Baltic Sea water for low concentration.

With the development of precise methods for measuring the electrical conductivity of sea water, a new definition of salinity based on conductivity rather than on chlorinity was recommended. The measurement of the conductivity of sea water is now becoming a standard procedure at sea. Recently, new tables connecting refractive index anomaly with salinity have been added to the oceanographic literature.

An accuracy of ± 0.02 ppm was obtained with the standard titration technique, and with the conductivity measurements technique used today this accuracy has been raised to ± 0.005 ppm.

It must be emphasized that the salinity as defined is not equivalent to the total salt content, but for its application in the equation of state of the sea-water system it is sufficient to assume that a linear relationship exists between them:

$$\text{Total salt content} = 0.043 + 1.0044 \text{ salinity}$$

In other words, salinity is a physical concept that indicates the concentration of the dissolved salts as a single solute. The high degree of accuracy required in its determination is due to the fact that the density of sea water has to be known to the fifth decimal place. This requirement must be fulfilled if the small horizontal pressure gradients that generate motion in the sea have to be determined. This will be explained in more detail in Chapter 4. Salinity has also been used in studies of chemical and biological nature. However, its use is restricted to the concept stated above, to indicate the concentration of the solution in the sea-water system for general correlation of different phenomena with the marine environment. For studies of more specific nature, the concentration of the particular ion has to be determined by the proper technique.

The origin of the salts dissolved in the ocean and the range of salinity have been mentioned before. It is useful to have some idea of the horizontal and vertical distribution of salinity in the world ocean, as well as its variations.

The greatest variations in the dissolved salts occur near the surface; in the deeper waters below 4000 m, the salinity values are generally between 34.6 and 34.8 ppm. The factors that affect the surface salinity of the oceans are listed in Table 3.1.

TABLE 3.1

Factors Increasing Salinity	Factors Decreasing Salinity
Evaporation	Precipitation
Ice formation	Ice melting
Surface circulation (advection of more saline water)	Surface circulation (advection of less saline water)
Mixing with more saline deep water	Mixing with less saline deep water
Solution of salt deposits (Gulf of Suez Canal, Gulf of Mexico)	Inflow of fresh water from the land (rivers, glaciers, icebergs)

Evaporation and precipitation are the main processes that change surface salinity. The propagation of these changes to the deeper layers is different for each. In the case of evaporation, the density of the surface water is increased by a twofold mechanism: the decrease in temperature owing to the loss of the heat by vaporization and the increase of salinity caused by the evaporation of fresh water. This creates a temporary vertical instability and a convective motion starts that tends to destroy the surface anomaly in temperature and salinity. Precipitation on the other hand adds fresh water at the surface and develops a strong positive stability which restricts the convective motion down so the temperature and salinity anomalies are not easily destroyed.

The strong correlation of surface salinity with the rates of evaporation and precipitation is clearly shown in Fig. 3.1, which shows mean values of evaporation and precipitation over the ocean from the Arctic to Antarctic according to Wust (1954), and also the comparison between the meridional variation of surface salinity with the corresponding excess of evaporation over precipitation. The values are averaged over season and longitude. There is some logical explanation for

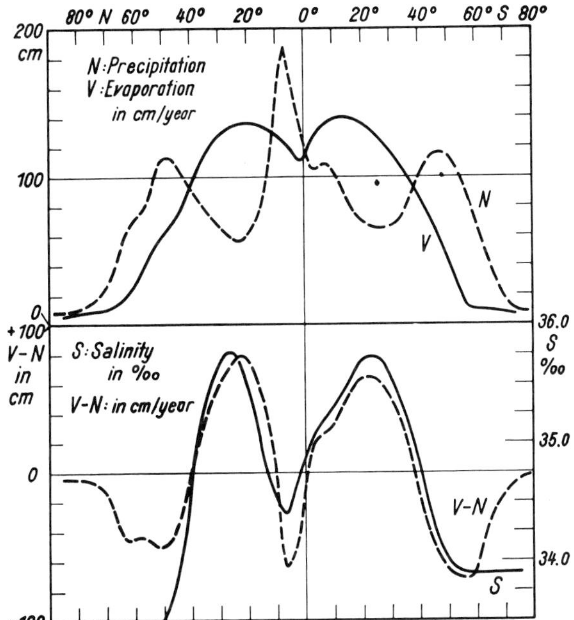

Fig. 3.1. Mean meridional distribution of precipitation N, evaporation V, $V - N$, and salinity at the ocean surface. The close relationship between the distributions of S and $V - N$ is clearly shown (according to Wüst, 1954). (From G. Dietrich and K. Kalle, *General Oceanography*, New York: Wiley-Interscience, 1953.)

the two salinity maxima near 30° N and 30° S as they are located near the latitudes of the atmospheric high-pressure cells over each ocean basin. The clear sky of these regions makes for considerable heating and consequent evaporation so that the maximum values of surface temperature and salinity tend to develop there. This is one typical example of the interaction between the oceans and the atmosphere.

The vertical distribution of salinity is characterized by the fact that the highest values are found at the surface or in the uppermost layers and usually decrease downward. It is not necessary for the salinity to increase with depth in order to maintain a stable density stratification in the ocean, as the temperature generally decreases with depth. Figure 3.2 shows the vertical distribution of salinity down to 4000 m for a series of oceanographic stations along a meridional section through

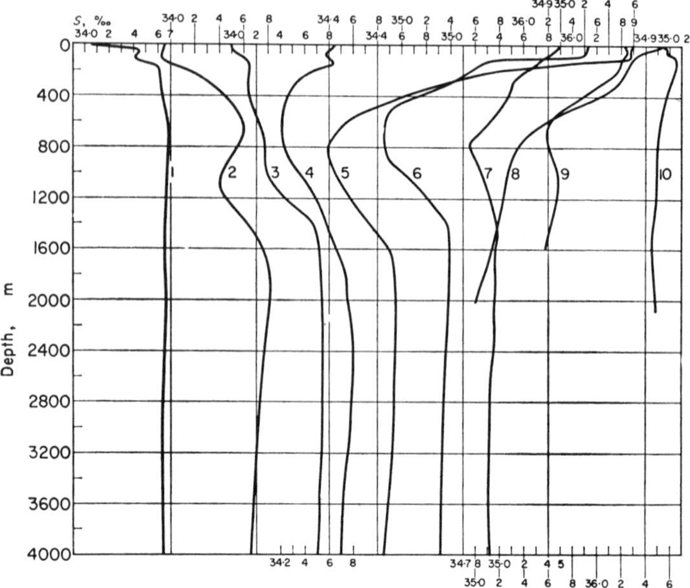

Fig. 3.2.   Vertical salinity curves for a series of oceanographic stations along a meridional section through the Atlantic. Corresponding vertical temperature curves are shown in Fig. 3.3. (From A. Defant, *Physical Oceanography*, Vol. I, New York: Pergamon Press, 1961.)

the Atlantic Ocean. This meridional cross-section extends from latitudes 63°S (station 1) to 78°N (station 10) through the central part of the Atlantic. The particular location of the oceanographic stations is of no interest because it is intended only to show the latitudinal variations of the vertical salinity distribution.

The propagation of salinity maxima and minima in the oceans has been a useful tool to study the spreadings and mixing of different water masses.

TEMPERATURE

The sea-water temperature is another of the important oceanographic elements that characterizes the water masses of the ocean. It is one of the variables that enters in the computation of the density of sea water and is measured with a high degree of accuracy ($\pm 0.02°$C) as was the case in salinity because of the precision requirements in the computation of density.

The greatest changes in temperature of the water take place at the surface, and the different processes which produce them are as follows: absorption of incoming short-wave solar radiation, emission of long-wave radiation from the sea surface back to the air, evaporation and condensation of water vapor, and convection of sensible heat to and from the atmosphere. The propagation of these disturbances in the vertical and horizontal directions depends on the values of the respective eddy coefficients.

The vertical temperature distribution in the ocean is such that the stratification, insofar as it depends on the temperature, is stable. In the oceanic troposphere, the decrease in temperature is so large that it compensates the vertical decrease in salinity and the equilibrium remains stable. In the upper layers of the stratosphere, the stratification is still stable; but at very great depths, below about 4500 m, particularly in the deep-sea basins, the vertical temperature distribution approaches the adiabatic and may even exceed it a little, so that there is an indifferent stratification or in some cases it may even be slightly unstable.

With the development of continuous measuring devices, it has been observed that the actual variation of temperature and salinity with depth is not smooth but goes in the typical steplike microstructure, which characterizes these fields on a small scale. Because of the influence of this phenomenon in underwater sound propagation, considerable research on the origin of this structure is presently being undertaken.

The vertical distribution of temperature in the ocean is shown by Fig. 3.3, which, although by no means typical, is representative of the temperature profile in the ocean. The characteristics of this vertical distribution of temperature are (1) a "mixed layer" or homogeneous warm surface layer which is produced by absorption of solar radiation and wind stirrings —its thickness ranges from tens of meters near the equator to more than 100 m in high latitudes and provides a fair indication of the degree of coupling with the atmosphere at the sea surface; (2) the *thermocline* region in which the temperature decreases rapidly with depth from the mixed layer above to the deep water below; and (3) the deep region in which the temperature falls very slowly with depth.

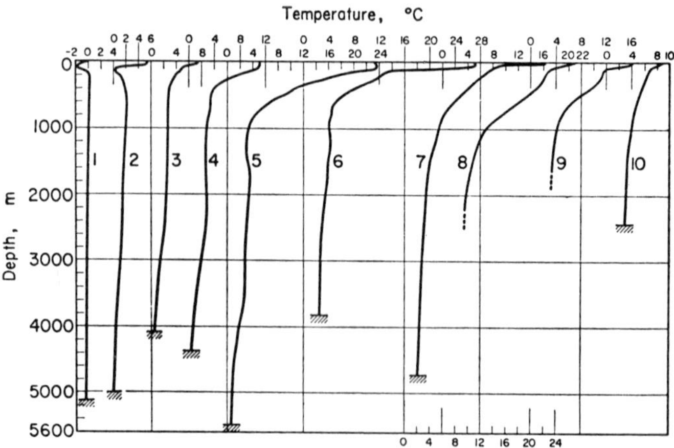

Fig. 3.3. Vertical temperature distribution at a series of stations along a meridian in the Atlantic Ocean. Corresponding vertical salinity curves are shown in Fig. 3.2. (From A. Defant, *Physical Oceanography*, Vol. I, New York: Pergamon Press, 1961.)

The *permanent thermocline* is the most striking characteristic of the earth's ocean and is generally centered at the 8 or 10°C isotherms whose depth range is from 300 to 400 m in equatorial regions to 500 to 1000 m in subtropical regions. This region of rapidly changing temperature is also a region of great stability. The term *thermocline* also applies to another region of abrupt change of temperature above the permanent thermocline and within the mixed layer. This shallower thermocline is referred to as *seasonal thermocline* and reflects the typical coupling between the air and sea at the particular season. There is also, in certain regions, a *diurnal thermocline*, commonly known as *afternoon effect*, which reflects the daily thermal processes taking place at the air–sea boundary layer.

The variation of the field of temperature in the ocean is also the object of considerable interest. Most of the classic literature deals with diurnal and annual variation of the temperature.

The diurnal variations of temperature in the open sea are usually smaller than 0.4°C and can rise at the most to about 1°C in calm and fair weather. They decrease rapidly with depth. As a whole, the individual daily temperature range is of only minor significance in the ocean, but it represents the shortest cycle of oceanic heat distribution.

Changes in temperature over longer periods can be investigated in two ways, namely to follow a definite water mass as it moves in the ocean and determine the individual time change of temperature or to select one particular fixed point in the ocean and study the local time change of temperature. For most parts of the world ocean, the annual displacements of the main ocean currents are known and the major annual changes in the sea temperature in those areas can be ascribed to these. For smaller areas, the annual variation of the sea surface can be derived only from a statistical evaluation of ships' observations, and for the deeper layers from series observations made by oceanographic expeditions.

In general, the annual temperature variation of the surface and upper layers of the sea depend on different factors, particularly on the annual variation in radiation, on the ocean currents, and on the prevailing winds. In equatorial and polar regions, the annual variation remains small, less than $2°C$; and the maximum value of approximately $7°C$ has been observed in latitude 30–40° where the radiation maximum is also found. In the western parts of the ocean, close to the continents and in small mediterranean and marginal seas which are under continental influences, larger sea surface temperature variations of up to $18°C$ occur.

In contrast to daily variation, the annual changes significantly influence the physical, chemical, and biological processes in the ocean.

DENSITY

The most striking phenomenon of the earth's atmosphere is the diminution of its density with altitude. This stratification endows it with a stability that is completely lacking in a homogeneous fluid. The earth's oceans are also stratified and have a stability of the same order as that of the atmosphere. This characteristic of our planetary fluids is perhaps responsible for the failure of classic hydrodynamics, which is concerned with the motion of a homogeneous, incompressible fluid, in explaining meteorological and oceanographic phenomena. The importance of the knowledge of mass or density fields in the ocean cannot be overemphasized.

The density of sea water is a function of its salt content or salinity, its temperature, and, to a lesser degree, its pressure $(\rho_{s,t,p})$. It is evident that its distribution in the ocean will depend on the distribution of temperature and salinity. The subscripts s, t, and p stand for actual salinity, temperature, and pressure.

The average density of sea water is close to 1.0276 g per cu cm. The significant part of this number with respect to the density of other water samples is generally in and beyond the third decimal. Therefore, a convention has been adopted that instead of the density $\rho_{s,t,p}$ a quantity $\sigma_{s,t,p}$ be used. The relation between the two is

$$\sigma_{s,t,p} = (\rho_{s,t,p} - 1)\,1000$$

Therefore, a density of 1.02542 g per cu cm becomes a sigma of 25.42 g per cu cm.

In physical oceanography, the density of sea water in situ has to be known to an accuracy of the fifth decimal place. This degree of precision is required to determine the small relative inclination of the isobaric surfaces in the ocean from which the field of motion will be derived. At the present time, there are no ways of direct measurements of density in situ with such a precision, so this quantity has to be computed from the equation of state of sea water which relies on empirical knowledge of thermal expansion of sea water, as well as saline contraction and isothermal compressibility. Because of nonlinear interactions, each of these is a somewhat imperfectly known function of the ambient temperature, salinity, and pressure.

The calculation of density is a tedious operation to be made by hand and requires the use of different tables. Modern computers have overcome this problem and the density at different depths or pressures is obtained in the normal processing of the oceanographic station data.

In considering the vertical distribution of density in the oceans, the use of a quantity *sigma tee* (σ_t) has been found very useful. This quantity is the density of the water that has been brought to the sea surface and differs by a very small amount from *potential density*. The quantity $\sigma_{s,t,p}$ is usually abbreviated σ_t and is very useful in stability consideration of the water column as it provides the information on whether a displaced parcel of water will be heavier or lighter than its surroundings from consideration of only its temperature and salinity. Also, the σ_t surfaces in the ocean can be considered nearly equivalent to the isentropic surfaces in a dry atmosphere and they are called quasi-isentropic surfaces. The name implies only that interchange or mixing of water masses along σ_t surfaces brings about small changes of the potential energy and of the entropy of the body of water.

The distribution of the density of the ocean waters is characterized by two features. In a vertical direction, the stratification is generally stable; and in a horizontal direction, differences in density can exist only in the presence of currents.

Since the density of sea water depends on its temperature and salinity, all processes that alter the temperature or the salinity influence the density. At sea surface, the density is increased by cooling, evaporation, and formation of ice, and is decreased by heating, condensation of water vapor, precipitation, meltwater from ice, and runoff from land. If the density of the surface water is increased beyond that of the underlying strata, vertical convection currents arise that lead to the formation of a layer of homogeneous water. Where intensive cooling, evaporation, or freezing takes place, the vertical convection currents penetrate to greater and greater depths until the density has attained a uniform value from the surface to the bottom. When this state has been reached, continued increase of the density of the surface water leads to an accumulation of the heaviest water near the bottom; and if the process continues in an area which is in free communication with other areas, this bottom water of great density spreads to other regions. Where deep or bottom water of great density is already present, the sinking water spreads at an intermediate level.

This is the basic reasoning to explain how the water masses are formed and fill the ocean basins. The point to bear in mind is that the waters of the oceans all attained their original characteristics when the water was in contact with the atmosphere, and then sank and spread along surfaces of similar density. During this process, some mixing takes place which alters the characteristics of the original water mass.

TEMPERATURE–SALINITY DIAGRAM

So far, the variation of temperature and salinity with depth has been studied. It is almost unconsciously assumed that these elements are independent of each other. This is, however, not the case. If salinity is plotted against temperature in a system of coordinates, the points for each depth are not distributed at random over the diagram but fall on a definite, more or less smooth curve. It is found that for oceanic regions with uniform oceanographic and climatic, as well as undisturbed flow conditions, the T–S relationship is quite characteristic. A given temperature corresponds to a given salinity regardless of the depth. When many separate observations of temperature and salinity of various geographic areas are plotted in this type of diagram, it is not uncommon for the points to fall in groups to the extent that oceanic areas and layers can be distinguished from one another as water masses. If any water mass is homogeneous, then the oceanographic elements in it are constant and it can be represented on the T–S diagram by a *single point*. This point defines a *water type*.

The T–S diagram has become one of the most valuable tools in physical oceanography. By means of this diagram, characteristic features of the temperature–salinity distribution are conveniently represented and anomalies in the distribution are easily recognized. It is also widely used in the analysis of oceanographic station data for detecting possible errors in the determination of temperature or salinity. It also has further uses in that with knowledge of the temperature and salinity of water at atmospheric pressure, σ, is also specified. At constant pressure, the density of sea water increases systematically with an increase of salinity and with a decrease in temperature. Therefore, in a T–S diagram, it is possible to rule a grid of lines of constant σ_t.

Figure 3.4 shows an example of a T–S curve for station No. 171 occupied by the research vessel *Meteor* in the central part of the South Atlantic. Its shape is characteristic for the entire South Atlantic from 40°S to beyond 10°N. In addition, lines of equal density σ_t (isopycnals) were drawn from which certain information about the vertical stability can be obtained. If the T–S curve of a certain layer runs approximately parallel

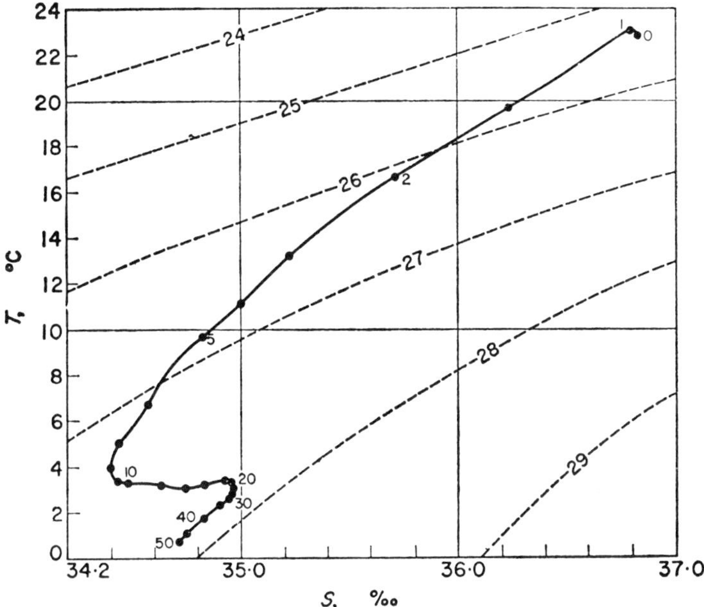

Fig. 3.4. *T-S* curve for *Meteor* station 171 (22° 1.5' S, 23° 47.0' W) in the central part of the South Atlantic (the thin dashed curves are the isopycnals σ_t). (From A. Defant, *Physical Oceanography*, Vol. I, New York: Pergamon Press, 1961.)

to the isopycnals, the stability in the layer is small but if the *T–S* curve cuts the isopycnals at a wide angle, the stability is larger.

The lines of equal σ_t shown in Fig. 3.4 are slightly curved downward and are inclined to the coordinate axes. Thus, when two water types of different salinity and temperature lie side by side on the ocean surface and have precisely the same density, their mixture will be represented by a point on the straight line joining their positions in the *T–S* diagram. Since this point must lie on the concave side of the σ_t line, the mixture must be slightly more dense than either of the two parent water types and tend to sink. Sinking caused by this mixing process is known as *caballing*. Caballing is presumed to take place at significant rates in high northern latitudes on the basis of the small changes of properties with depth found, which perhaps bears testimony to the effectiveness of the process.

ADIABATIC PROCESSES

An *adiabatic* process is defined as one in which there is no heat exchange between the system and the environment. Therefore, when a portion of a fluid is compressed or expanded without exchange of heat with the surrounding fluid, there are temperature changes. Such adiabatic temperature changes are very important in the atmosphere and are determined solely by pressure changes. In the oceans, the effects are less important, although because of the compressibility of the sea water the temperature varies slightly with the amount of compression. The compressibility of water is less than that of steel; however, as the depth of the water and the hydrostatic pressure increase, the temperature of the water also increases. The rate of increase is approximately 0.03°C per 1000 m or decibars increase of pressure.

Adiabatic processes are also known as isentropic processes because of the constancy of the entropy of the system in any reversible adiabatic cycle. The study of atmospheric processes along surfaces of constant entropy has been widely used in meteorology and is known as the *isentropic analysis* method.

Two concepts which are useful in the analysis of deep water structure in the ocean have been obtained from the equation of entropy change, namely the *adiabatic lapse rate of temperature* and *potential temperature.*

The adiabatic lapse rate of temperature depends mainly on the coefficient of thermal expansion, which varies with temperature and pressure. Ekman has computed the adiabatic temperature gradient for different temperatures, salinities, and depths.

Potential temperature is defined as the temperature which a parcel of water of certain temperature and salinity at a certain depth will attain if raised adiabatically to the sea surface. The potential temperature is useful in discussing the movement of deep water in the ocean because the effects of pressure on temperature are removed. Thus, temperatures of the water at different depths may be compared.

In the very deepest parts of the oceans, such as in trenches flanking island arcs, the actual temperature increases somewhat with depth. This observed positive gradient may suggest an apparent instability. However, it can be shown that the potential temperature of the water at those depths is virtually uniform from the depth of the surrounding ocean floor to the bottom of the trench, suggesting a convective overturning in the trench. Adiabatic warming and cooling due to vertical motions of sea water have been observed in relatively deep and confined spaces where the destroying effects of weak advection can be suppressed by the confinement of the region.

Adiabatic effects are also considered in the propagation of compressional sound waves in the ocean as the instantaneous pressure changes of the sound waves create adiabatic changes in the perturbated water.

STRATIFICATION

The important characteristics of the stratification of oceanic waters have been stated. Owing to the action of gravity and buoyant forces, there is a tendency for dense parcels of water to sink and for less dense parcels to rise toward the surface of the sea. The final result of this convective motion is the establishment of a more or less stable vertical gradient of density. As density is a function of temperature, salinity, and pressure, the stably stratified water column may often be cooler at the bottom than at the top, but also it may be less saline at the bottom. This particular dependence has to be kept in mind as the balance of the opposing effects of increasing salinity and increasing temperature would apparently result in unstable configurations. Rather, what results is a stable stratified column from the standpoint of the vertical distribution of density.

The vertical distribution of temperature in the ocean has been discussed. Particularly stressed were the occurrences of steep vertical gradients of temperature of the main or permanent thermocline deep enough to be affected by seasonal changes at the surface and the shallower seasonal thermocline that comes and goes with the seasons. Along with the temperature transition in the thermoclines, there may be locally steepened vertical gradients of salinity and density, referred to as *haloclines* and *pycnoclines*, respectively. The physical processes which originate and maintain these transition layers are thought to be associated with a balance of the downward diffusion of heat and the salt-enriched surface layers against the upward motion of fresher and cooler water from the ocean depths. The presence of the permanent thermocline is related to the problem of the general circulation of the oceans.

The vertical stratification of the water in the ocean may occur in two different forms: *barotropic* or *baroclinic stratification*.

In a barotropic ocean, Fig. 3.5, the density is a function of pressure alone and the consequence is that the isopycnal and the isobaric surfaces are parallel. They may or may not be parallel

BAROTROPIC

Density, ρ, is a function of pressure, p, alone, hence isobaric and isopycnal surfaces do not intersect. The fluid can be motionless when isobaric surfaces coincide with geopotential surfaces.

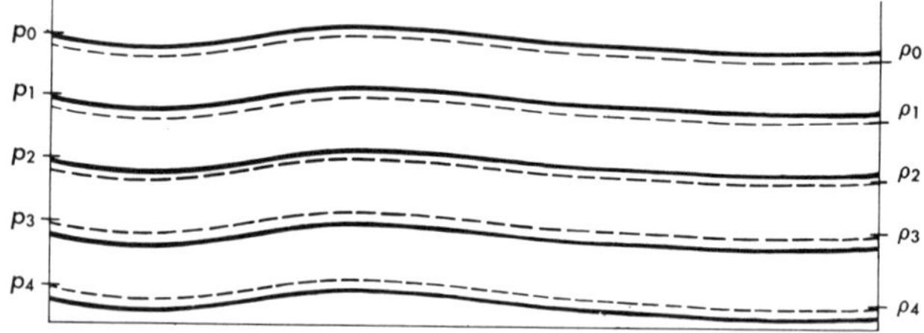

BAROCLINIC MODE

Isobaric and isopycnal surfaces intersect. A baroclinic fluid cannot remain motionless.

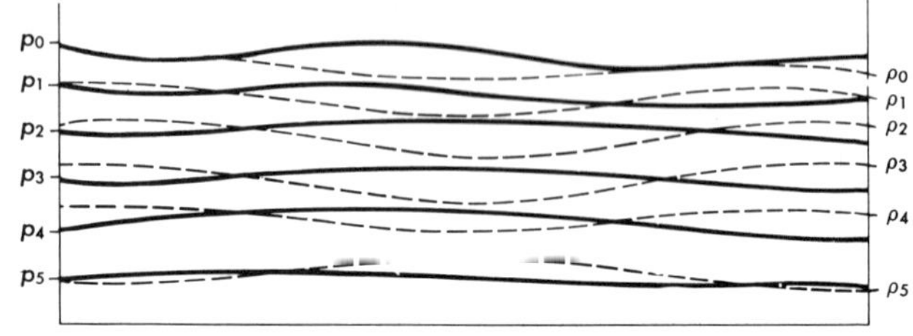

Fig. 3.5. Ocean stratification. (From W. S. Von Arx, *Introduction to Physical Oceanography*, Reading, Mass.: Addison-Wesley, 1962.)

to the geopotential surfaces depending on the presence of the barotropic motion. When a barotropic ocean is at rest, the isobaric and isopycnal surfaces are parallel with geopotential surfaces. But when there is geostrophic motion accompanying the barotropic condition, the isobaric and isopycnal surfaces, provided that they remain parallel, may be inclined to geopotential surfaces. Barotropic stratification is not destroyed when acted upon by such fields of force as the gravitational attraction producing the astronomic tide and on a smaller scale those accompanying a widespread change of barometric pressure. Those fields act bodily in the whole water column and the whole field of pressure is affected similarly. In cases where such force fields are suddenly applied (as the passing of an atmospheric front) and some redistribution of mass is required to achieve equilibrium, the period of readjustment is virtually complete in 12 pendulum hr. Where, however, the acting field of force establishes steady patterns of circulation, such as the large seasonal variations of winds and the air motions around the semipermanent centers of high and low pressure over the oceans, the period of barotropic equilibrium is of the order of 12 to 14 days in middle latitudes.

Baroclinic stratification, Fig. 3.5, exists when isobaric and isopycnal surfaces intersect. Both may be inclined to geopotential surfaces. The situation is illustrated by the unequal stands of liquid level in a U-tube when the two arms contain a succession of fluids of different densities. Baroclinic stratifications are especially pronounced in the upper 500 to 1000 m of the ocean. At greater depths, the conditions approach barotropy. Where there is a sustained baroclinity, there is also water motion because of the inclination of the isobaric surfaces. The baroclinic structure which is produced by regional inequalities of density is usually maintained by persistent external influences such as climate; and if baroclinity persists, the associated flow must be geostrophic. Because of the great volumes of water in the major oceanic regions of the earth, it takes considerable time—tens or even hundreds of years—for appreciable changes to take place in the baroclinic stratification already existing. This is principally because sea water is not easily penetrated by sunlight and heat, and yet it has an enormous capacity for heat storage. The coupling of the ocean and the atmosphere is such that the ocean tries to get in phase with the changes of atmospheric pressure and wind stress by modifying the barotropic part of its circulation pattern, which reacts quickly to the perturbation. However, as the baroclinic process requires a major reorganization of the fields of temperature and salinity, it is not possible to reach an equilibrium with the seasonal wind field before changing seasons. The oceans are constantly in a state of readjustment with respect to the winds because of the long time required by the baroclinic process to get adjusted to the variations.

The degree of stability of a fluid is normally indicated through a stability coefficient N known as the *Väisälä-Brunt frequency*. The best way to visualize this stability frequency is by consideration of the processes affecting a parcel of water which is moved up or down from its equilibrium position in a fluid of stable stratification. The buoyant force moves the parcel back to its original position where it rebounds from any displacement in a kind of oscillatory motion similar to the motion of a spring suddenly released from tension. The frequency at which this occurs, the Väisälä-Brunt frequency, may be computed from the density structure.

A typical vertical distribution of the stability frequency in the ocean shows a maximum near the surface which corresponds to the seasonal thermocline. A smaller maximum between 800 and 1000 m is related to the permanent thermocline, and below the thermocline there is a continuous decrease with depth. The values of the stability period for the seasonal thermocline are around 13 min the permanent thermocline around 21 min, and the deep waters around $2\frac{1}{2}$ hr.

HEAT CAPACITY

By far the largest source of energy input in the ocean is the solar radiant energy, and 90 percent of this is absorbed in the upper 40 m of water. Both the near- and especially the far-infrared bands of solar radiation are rapidly absorbed in the first few millimeters of water depth. Visible light is more gradually transformed into heat because it penetrates far more deeply into the sea before being absorbed. In the presence of turbulence produced by waves and currents, both the shallow and deep radiant heating of the sea is mixed downward within a few hours to a depth of around 100 m under most circumstances.

The best way to show the great heat capacity of the ocean is to compare it with the same property on land. Assuming that solar radiation is absorbed at the same rate on land and sea, the rise in temperature per unit time of a cubic centimeter of water at the sea surface can be calculated and compared with the rise in temperature of a cubic centimeter of bare rock under like conditions. If 1 g-cal per min per sq cm reaches the surface of the earth, it is found that in the rock, if it is heated without reflection to a depth of 1 cm, the rate of temperature rise is $1°C$ per min, which is not strikingly less until it is considered that the sea mixes its heat downward. The first centimeter of heated sea water, therefore, shares its heat content with a water column perhaps 100 m deep. Thus, the average temperature rise in the sea surface is $10^{-4}°C$ per min. From this it is concluded that the oceans are warmed very slowly and, compared with land, can accommodate a very large amount of heat at a given surface temperature because of vertical mixing. Only under special conditions when vertical mixing is not strong enough because of a very calm day, a shallow thermocline develops, and the afternoon sea surface temperature rises a degree or two above the temperature of the water a few meters below the surface.

There is a constant inflow of radiant energy from the sun and a constant outgoing radiation from the earth into space. In the oceans, there also exists a quasistationary state insofar as heat energy is concerned. Although there are small variations with time in the temperature of the ocean waters, these can be taken as variations around a "mean" value which remains essentially unchanged. In this quasistationary state, all the supply in energy is balanced by equally large losses of energy. The most important processes in this dynamic equilibrium are the radiation, the interchange of sensible heat with the atmosphere above the sea, and evaporation from the sea surface or the condensation of atmospheric water vapor. These processes as well as other sources and sinks of heat and the estimated average fluxes of thermal energy are shown in Fig. 3.6.

OPTICAL PROPERTIES

Optical properties of sea water influence not only the intensity and spectral composition of the light that is available for the photosynthetic process, but also the visual conditions in the ocean and the color of the sea water. Sunlight incident upon a glassy smooth ocean is partly reflected and the balance is refracted, scattered, or absorbed. The amount of visible light re-

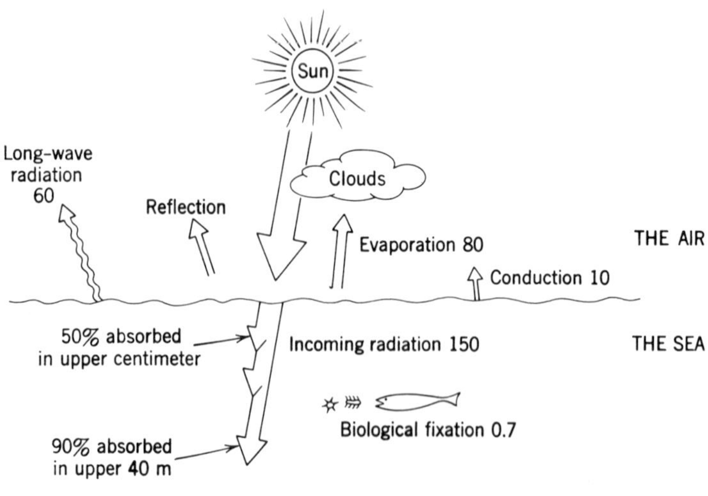

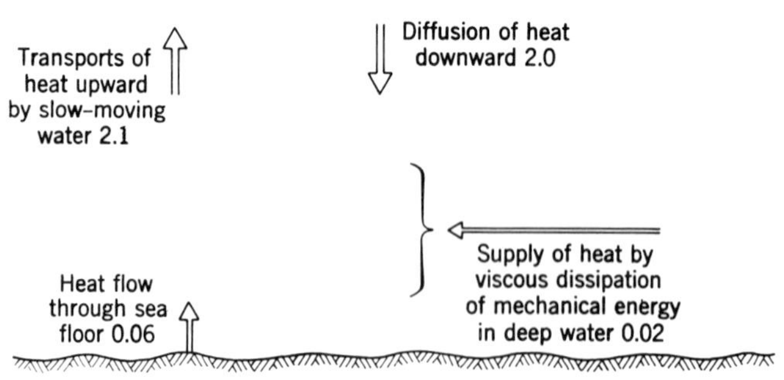

Fig. 3.6. Estimated average fluxes of thermal energy. All figures represent energy flow in watts per square meter of sea surface. (From J. F. Brahtz, *Ocean Engineering*, New York: Wiley, 1968.)

flected varies with the angle of incidence, a property that is not shared by the rocky, grassy, or forest-covered areas of the earth. However, in the normally ruffled or wavy condition of the sea, the angle of incidence is highly varied; and, thus, absorption continues during the day unless foam is sufficiently abundant to increase the reflection coefficient of the sea surface.

The portion that is not reflected is refracted downward toward the normal in accordance with Snell's law. Thus, viewed from beneath a glassy smooth sea surface, the sun appears closer to the vertical than it appears when viewed above the sea surface. Going deeper in calm water, the intensity of light diminishes owing to attenuation through scattering and absorption in the water. The effect of these diffusing processes leads to a change in the direction of maximum radiance: from the direction of the refracted solar beam toward the vertical as the depth is increased. At the same time, there is absorption primarily of red light which causes the visual field to appear more and more blue-green in color. With increasing depth, the scattering process causes the upward flux of light to contribute a large fraction of the total radiance reaching a point so that when looking down, the water appears to glow.

The reduction in intensity of the radiation is often characterized in practice by the extinction coefficient which, as indicated above, is a function of the wavelength. This coefficient takes account of the effects of both scattering and absorption. Figure 3.7 shows the values of the extinction coefficients of radiation at

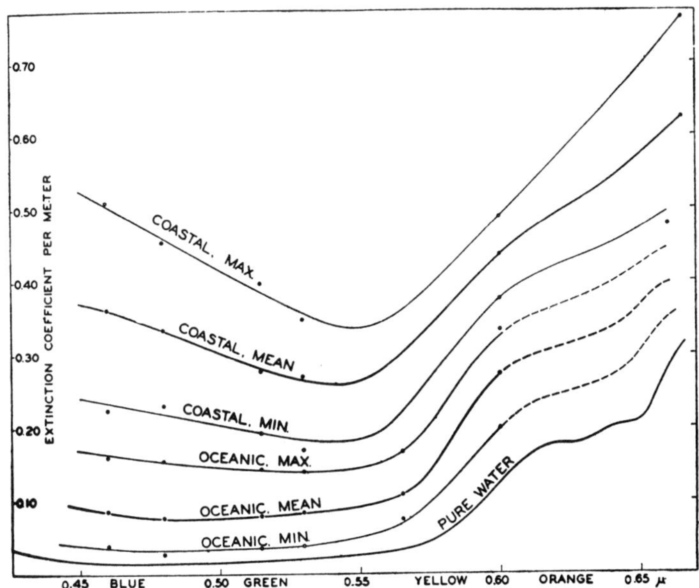

Fig. 3.7. Extinction coefficients of radiation of different wavelengths in pure water and in different types of sea water. (From H. U. Sverdrup, M. W. Johnson, and R. H. Fleming, *The Oceans*, Englewood Cliffs, N.J.: Prentice-Hall, 1942.)

different wavelengths in different kinds of water. The great values in the sea, as compared to those of absolutely pure water, are, as a rule, ascribed to the presence of minute particles which cause scattering and reflection of the radiation and which, themselves, absorb radiation.

ACOUSTICAL PROPERTIES

One important characteristic of the ocean environment is the opacity to electromagnetic energy, and the high conductivity of sounds. The high attenuation and scattering of light energy shown before is typical of how the marine environment reacts with electromagnetic energy.

The propagation of sound in the ocean is of particular interest since the acoustical soundings and range determinations, which have so many applications in oceanography, fishing, and maritime navigation, are based on a knowledge of sound propagation in sea water. Velocity, refraction, and absorption of sound determine the way it propagates in the marine environment.

Water is a very efficient medium for the transmission of sound, which travels more rapidly and with much less absorption of energy through water than through air. Sound waves, like all other elastic waves, exist by virtue of a restoring response of the medium to disturbance. The finite compressibility of sea water provides the restoring force.

The velocity of sound in sea water is independent of the wavelength, except for impulses resulting from the detonation of relatively large amounts of explosives. This velocity is a function of the elasticity and density of the medium which change with temperature, salinity, and pressure. In sea water of normal salinity, the speed of sound is primarily a function of temperature and pressure. The speed decreases as the temperature is decreased, but increases with increasing pressure. In the open ocean where salinity changes are very small, the temperature profile normally present, combined with the hydrostatic pressure, leads to a vertical distribution of sound velocity with a pronounced minimum at some intermediate depth, which varies from around 1000 m in low and middle latitudes to practically the surface in polar regions. Figure 3.8 shows some typical

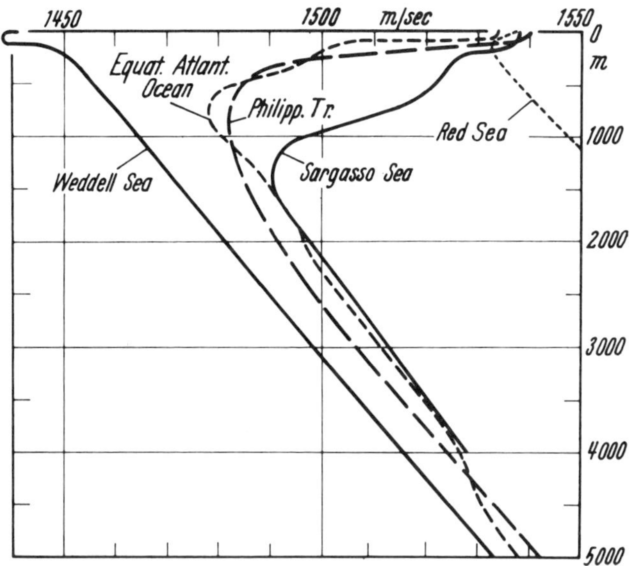

Fig. 3.8. Examples of horizontal velocity of sound in the world ocean (according to G. Dietrich, 1952). (From G. Dietrich and K. Kalle, *General Oceanography*, New York: Wiley-Interscience, 1953.)

vertical distributions of sound velocity for different seas of the world ocean.

In deep water, sound may travel from source to receiver by surface ducts, sound channel, convergence zone, and bottom bounce. The most useful one is the sound channel whose axis lies at the depth of the velocity minimum. Use of this channel by the so-called SOFAR (sound fixing and ranging) technique is being used more widely for different purposes. The most remarkable characteristic of this channel is that very little acoustic energy is lost from it. The detonation of several kilograms of explosive in the SOFAR channel generates a signal that propagates and can be detected several thousands of miles away by a hydrophone at the channel depth.

Analogous to optics where one material is called optically denser or thinner than another depending on whether the velocity of light in this substance is smaller or greater than in the other one, respectively, one also distinguishes between an acoustically denser or thinner medium if the velocity behaves similarly. The propagation of sound in a stratified oceanic medium follows Snell's law of refraction, as was the case with radiant energy. Figure 3.9 shows different sound transmission paths that correspond to different profiles of sound velocity.

Sound energy propagated through a volume of sea water

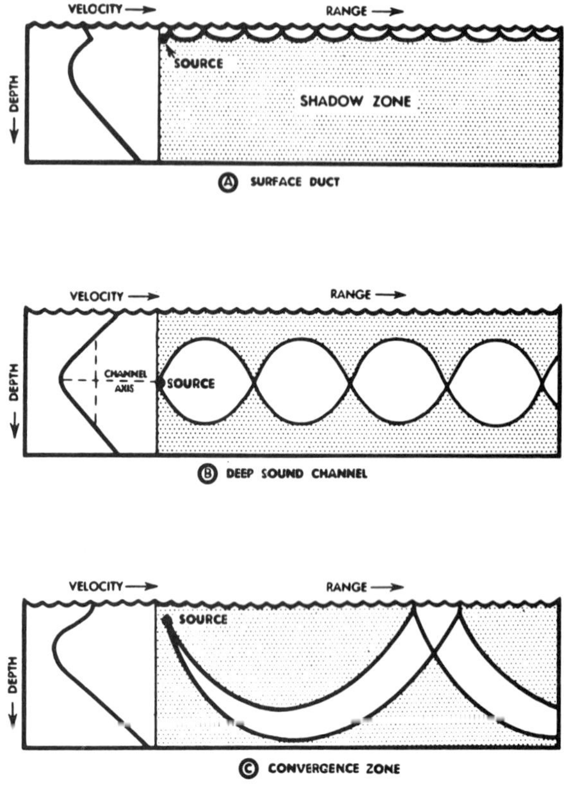

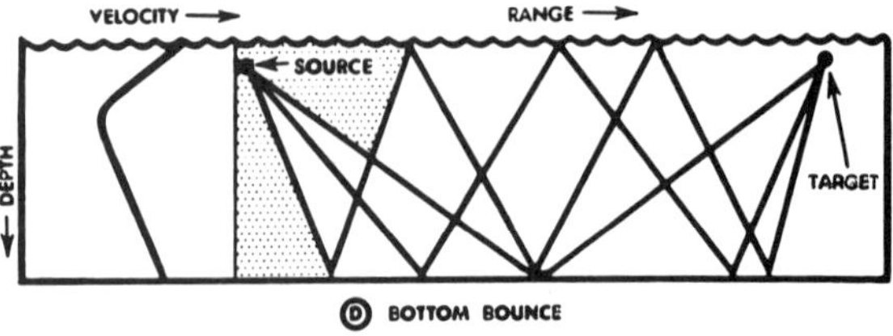

Fig. 3.9.   Sound transmission paths.

undergoes some loss of energy because of attenuation, that is, absorption and scattering. In the passage of sound through water, some of the energy is converted into heat; this is called absorption. Scattering loss results from reflectors in the water that may vary in size from minute air bubbles to a whale. The normal scatterers in the marine environment are bubbles, marine organisms, irregularities of sea floor and sea surface, and density variations resulting from turbulence or internal waves.

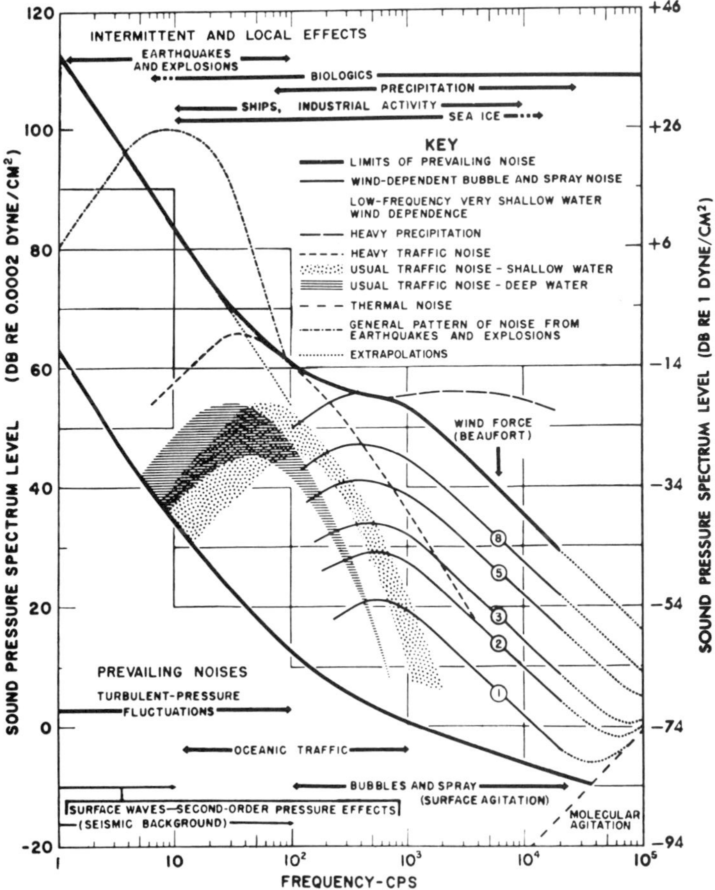

Fig. 3.10. A summary of ambient noise sources. (According to G. M. Wenz, *J. Acoust. Soc. Am.*, **34**, 1936, 1962.)

In the deep ocean, it is very common to record the presence of one or more layers of scatterers at different depths. These are known as the *deep scattering layers*, a phenomenon which puzzled scientists for many years. In general, there are three scattering layers lying between about 300 and 700 m (900 and 2000 ft) with variable thickness of 13 to 20 m (40 to 60 ft). Because these layers migrate vertically twice per day—rising to the surface at night and descending at sunrise—they are believed to consist of small marine organisms which rise at night to graze in the plankton-rich surface waters.

The use of listening devices in the ocean is often limited by the natural ambient noise, which is the noise inherent in the sea itself. Such ambient noise is most intense at the lower frequencies. Some characteristics of ambient noise sources are given in Fig. 3.10.

SUMMARY

Water substance has unique properties that, in many ways, condition the terrestrial environment. These anomalous properties are explained as the tendency of the water molecules in natural conditions to cluster in packages of two or three—polymerization—and the presence of heavy isotopes of hydrogen and oxygen. The presence of salts modifies the behavior very little and adds the properties of colligation common to solutions. Moreover, sea water, as the main constituent of a large-scale natural system, as it is in the world ocean, is in permanent motion and has a considerable amount of particles in suspension. These two particular characteristics greatly affect behavior and many of the properties. The state of motion which takes place in all scales and frequencies imparts to the marine environment the particularity of a turbulent fluid which destroys sharp gradients in concentrations and is permanently mixing the water layers. The suspended particles reduce considerably the amount of radiant energy that water can absorb and so affect the heat budget of the ocean.

The layered structure of the sea—stable stratification—is one of the striking characteristics of the marine environment. This particular characteristic has profound implications in ocean phenomena. It confines the processes that take place in the air-sea boundary layer, such as the transfer of energy and matter, absorption of solar radiation, and most of the biological activity of the sea to the upper layers of the ocean. The depth of the euphotic layer is related to the stratification of the ocean.

The equation of state of sea water is responsible for one important type of ocean stratification. The rate of decrease of the specific volume of water is a function of temperature. This characteristic is responsible for the isoteric surfaces sloping with respect to the isobaric surfaces. This crossing of surfaces imparts the baroclinic stratification, which is synonymous with water motion, to the marine environment. The other mode of stratification is much simpler, as both surfaces are parallel. The first type of stratification is common in the oceanic troposphere above the main thermocline and the second type is common in the deep ocean.

The genetic characteristics of the water masses, such as salinity, temperature, and dissolved gases, are acquired at the surface of the ocean by interaction with the atmosphere. Sea water of any depth has been conditioned somewhere, sometime, at the surface. Sea water is highly selective on the visible part of the spectrum. Certain frequencies are absorbed in the upper centimeters of the sea surface while others penetrate deeper, although they are badly scattered.

Of particular relevance for the engineer are the following:

1. The degree of polymerization and the abundance of heavy isotopes can make some areas of the ocean more valuable than others, considering some of the actual uses of sea water.

2. The typical distribution of density in the ocean is such that the most rigorous and variable environmental conditions will be found in the upper layers of the sea, particularly above the seasonal thermocline as the coupling with violent atmospheric phenomena do not generally go deeper than the thermocline.

3. At density discontinuities, important oscillations of the water layers take place. They are known as internal waves.

4. The density stratification is also responsible for poor acoustic behavior. On the other hand, the sound channel (SOFAR channel) which propagates acoustic vibrations to considerable distances, is developed because of the vertical distribution of density.

4
Ocean Circulation

The central problem of physical oceanography is the general circulation of the ocean. Broadly speaking, the observed oceanic circulation can be satisfactorily accounted for by the analysis of the two important modes of large-scale motion: advective circulation, commonly known as wind-driven circulation, and the convective circulation, known as thermohaline circulation. The former is representative of the major global systems of surface and subsurface currents found in the world ocean. The latter is related to the abyssal circulation.

The inclination of the axis of rotation of the earth with respect to the plane of the earth's orbit around the sun is responsible for the bulk of solar radiation falling near the earth's equator and creating a differential heating on the surface of the earth. The rotation of the earth distributes this energy zonally (east-west). The earth radiates long-wave radiation back into space, and the processes are such that in equatorial regions there is a surplus of radiation received by the sun while in higher latitudes there is a deficit. The temperature differences so generated require such an amount of heat to be transported toward the polar regions that no conductive processes can provide it. This task is taken by geophysical fluid motions or, more specifically, by winds and ocean currents.

There is, however, an important difference in the way atmospheric and oceanic circulations are generated. The air is almost transparent to solar radiation and receives its heat mainly from the sea surface and land so that atmospheric circulations are mainly generated by heat from below; the oceans, on the other hand, are heated from above. If the normal stratification of both media are considered, this basic difference in the manner that both media are fueled is responsible for the winds being more effective as a heat transport agent. The strong stability created in the surface layers of the oceans—because they are heated from above—reduces the effectiveness of planetary temperature contrasts as a main source of energy for advective circulation of the oceans. However, the pattern of the major global systems of surface currents in the ocean is almost identical to the pattern of the surface winds, a fact that indicates that the drag of the wind over the sea surface is the principal source of energy for the horizontal motions in the ocean. Through this mechanical interaction between the atmosphere and the ocean and perhaps also by regional excesses of evaporation over precipitation, the advective modes of ocean circulation are made sensitively responsive to the distribution of solar heat.

The wind-driven circulation has been the subject of considerable interest in physical oceanography and different investigators have produced different theories that explain the general picture of the circulation and some of its peculiar characteristics, particularly the intense crowding of streamlines near the western borders of the oceans so well exemplified by the Gulf Stream (westward intensification of ocean currents). Although the vorticity model of wind-driven currents deals successfully with the explanation of the swift narrow currents in the western parts of the ocean and with the concept of a slow and broad drift toward the equator on the eastern side to close the orbit of circulation in the wind-driven layer, it does not fit with the observed distribution of water properties along the surface route. This incompatibility and other minor objections have been met by the modern concept of the convective or thermohaline circulation.

The main driving force in this convective circulation is provided by the thermal and evaporation–precipitation balances. Again, as in the case of the wind-driven circulation, intensified deep circulations adjacent to the western boundaries of the ocean were found by dynamic computations and by direct measurements. The modern convective model that accounts for the observed phenomena and is compatible with the wind-driven circulation has been developed notably by Stommel and collaborators (1957), who based their work on the topography of the oceanic thermocline. This main thermocline provides a floor at well-known levels beneath the wind-driven circulation and a ceiling above the circulation of water in the abyss. In a remarkable study, Stommel explains the thermohaline circulation as a dynamic system consisting of highly localized sources of sinking water at high latitudes, which gives rise to a system of narrow deep currents flowing along the western boundary of the oceans (western boundary currents) that transfer the sinking water into the ocean basins. With change of latitude, which implies an increase of the Coriolis factor, geostrophic (mainly zonal) flows develop which supply the gentle upward flow required to support the main thermocline (distributed sink).

The existing knowledge of ocean circulation was obtained principally from computation of the field of motion from the observed distribution of density in the ocean. Theoretical, simplified models have been devised to explain the observed phenomena and some experimental laboratory studies have been carried out. Every one of these approaches has its shortcomings, and it can be said that the knowledge of the general circulation in the ocean is still quite inadequate. Thus, whereas a general picture of a very simple surface circulation has accumulated, the introduction of new techniques or the careful use of the old ones has produced disturbing evidence that the real ocean behaves quite differently from the simple scheme devised. Direct measurements below the surface using drogues, current meters, and neutrally buoyant floats have shown higher speeds and greater variability than previously suspected.

The difficulties seem to be related to an ignorance of the spectrum of motion in the sea. This motion appears to cover a considerable range of geometric dimensions and frequencies which are not necessarily those of principal importance in the ocean. Thus, the classic oceanographic station is inadequate to sample high-frequency and small-scale phenomena, and the duration and geographical extent of oceanographic expeditions are inadequate to sample low-frequency or large-dimensional phenomena. Continuous observations in time and space over vast areas of the ocean and for very long periods of time are desirable.

Before going into a more detailed discussion of the advective and convective modes of oceanic circulation, it is convenient to survey the nature and magnitude of the driving forces in the ocean as well as the hydrodynamic equations that describe the motion generated by those forces. These forces and equations are discussed in textbooks on hydrodynamics, but what is emphasized here is their applications in natural environments and geophysical scale.

FIELD OF FORCE

Gravity   Fluid motion can be initiated by very small forces. In the water of the ocean, it is usually a small component of the earth's gravitational field that provides the motive force whenever static equilibrium is disturbed directly or by solar heating. In general, the basic cause of the perturbation is overlooked in order to avoid the complicated problem of considering the totality of effects and interactions, and the simplification consists in measuring pressure differences from place to place on the ocean and, from Newtonian principles, deducing the resulting kinds of motion. Therefore, the horizontal pressure gradients, although extremely small, are all-important to the state of motion, whereas the vertical ones are insignificant in this respect. The horizontal pressure gradients vanish in a state of perfect hydrostatic equilibrium. Such a state would be present if the atmospheric pressure, acting on the sea surface, were constant, if the sea surface coincided with the ideal sea level, and if the density of the water depended on pressure only. This is not the case in the ocean where an internal field of force is generally present.

The common way to define this field of force is by considering the slope of the isobaric surfaces instead of the horizontal pressure gradients. As the isobaric surfaces will be inclined with respect to the level surfaces, a component of the acceleration of gravity acts along the isobaric surface and will tend to set the water in motion.

The important forces which drive the large-scale oceanic motions are the force of gravity, the pressure gradient force, the Coriolis force, and frictional forces.

Gravity is the result of the force of attraction of earthly masses and of the centrifugal force of the earth's rotation. Its distribution at the surface of the earth can be found by pendulum measurements.

The acceleration of gravity varies with latitude and depth, and because the gravitational force represents one of the most important of the acting forces, it is very convenient to use in problems of statics or dynamics a coordinate system where gravity could be held constant. With this purpose in mind, it was found convenient to define a new system of coordinate surfaces known as *level surfaces* or *surfaces that are everywhere normal to the force of gravity.* These surfaces do not coincide with surfaces of equal geometric depth.

It follows from the definition of level surfaces that, if no forces other than gravitation are acting, a mass can be moved along a level surface without expenditure of work and that the amount of work expended or gained by moving a unit of mass from one surface to another is independent of the path taken. The work required or gained in moving a unit of mass from sea level to a point above or below sea level is called the *gravity potential* and the practical unit to express it is the *dynamic meter.* This unit is larger by about 2 percent than the standard meter.

The degree of variation of the acceleration of gravity with latitude is clearly shown by the fact that at the North Pole, the geometric depth of the 1000 dynamic-meter surface is 1017.0 m; but at the equator, the depth is 1022.3 m because gravity is greater at the poles than at the equator. Thus, level surfaces slope relative to the surfaces of equal geometric depth, and, therefore, a component of the acceleration of gravity acts along surfaces of equal geometric depth. With any scalar field such as temperature, specific volume, or density, sea bottom depth can be represented by means of a series of topographic charts of equiscalar surfaces in which the contour lines represent the intersection of the level surfaces with the equiscalar surface. Charts of this character will be called geopotential topographic charts, or charts of *geopotential topography*, in contrast to topographic charts in which the contour lines represent the lines along which the depth of the surface under consideration is constant. The field of gravity in the ocean can be completely described by means of a set of equipotential surfaces corresponding to standard intervals of the gravity potential. Geopotential surfaces are level and a fluid particle on any of those surfaces will show no tendency to move unless acted on by some unbalanced force. Such forces may arise from thermal disturbances of hydrostatic equilibrium which tends to be restored by motions developed under the earth's gravity or, in the case of tides, from the action of the external gravitational field of the sun and moon.

Pressure The internal stress in a liquid such as the ocean is characterized by the pressure per unit area. In a liquid in equilibrium, owing to the absence of any resistance to deformation (which is typical of fluids) this pressure acts perpendicular to any arbitrarily oriented surface through the liquid and is equal for any point and in all directions. This state is denoted as hydrostatic stress state, and the pressure at every level is assumed to be exactly equal to the total weight of the fluid per unit area.

The pressure of sea water increases by 1 atmosphere (atm) over approximately 10 m (33 ft) of depth. The units of pressure usually employed in oceanography and meteorology are related to and are derived from the metric standard of 1 dyne per sq cm (barye). Because the mean atmospheric pressure is close to 10^6 dynes per sq cm, it is common to express atmospheric pressures in terms of the bar, which is defined as equivalent as 10^6 dynes per sq cm. In the ocean, it is convenient to measure ocean pressure in decibars and because either the density or the specific volume differ little from unity, a difference in pressure is expressed in decibars by nearly the same number that expresses the difference in geopotential in dynamic meters, or the difference in geometric depth in meters.

The pressure field can be completely described by a series of charts showing isobars at standard level surfaces or by a series of charts showing the geopotential topography of standard isobaric surfaces. The latter is commonly used in oceanography.

The increase of pressure with depth in the ocean is associated with a vertical *pressure gradient force* which is directed in the sense of decreasing pressures (upward) and normal to the isobaric surfaces. This pressure gradient has two principal components: the vertical, directed normally to the level surfaces, and the horizontal, directed parallel to the level surfaces. When static equilibrium exists, the vertical component is balanced by gravity. In a resting system, the horizontal component of the pressure gradient is not balanced by any other force; therefore the existence of a horizontal pressure gradient indicates that the system *is not at rest* or *cannot remain at rest*. It is again emphasized that the existence of the horizontal pressure gradients, although extremely small, are important to the state of motion.

It is evident that if one could determine the shape of the field of pressure in the ocean, the associated field of motion would be known. However, it is not yet feasible to measure the pressure in the ocean accurately enough for the needs of the observed circulation. In middle latitudes even the strongest surface currents rarely have velocities above 100 cm per sec (about 2 knots) and the corresponding slope of the isobaric surface is around 1 cm in 1 km. Such a gentle slope cannot possibly be observed directly. To overcome this limitation, oceanographers reverted to the indirect method known as the determination of the relative field of pressure. Basically, it consists of determining the slope of one isobaric surface relative to another through the knowledge of the density of the water at different points and depths. Therefore, if the field of mass or density in the ocean is known, and the sea surface is assumed to be an isobaric surface, the slope of this sea surface relative to an isobaric surface at any depth can be determined. One example will clarify the method. If at two neighboring stations, *A* and *B*, the average density between the sea surface and 1500 m is less at *A* than it is at *B*, the distance between an isobaric surface at a nominal depth of 1500 m and the sea surface is greater at *A* than it is at *B*. Relative to the isobaric surface at about 1500 m, the sea surface slopes downward from *A* to *B*. If this relative slope is to be determined with sufficient accuracy, the density must be known to the fifth decimal place. As the density at any particular depth is computed from the temperature and salinity, this means that errors in temperature and salinity must not exceed 0.02°C and 0.02 ppm, respectively. This explains the many efforts to improve the accuracy of salinity and temperature measurements.

From relative slopes *relative currents* can be computed, but the aim is to determine the *absolute currents* so it is necessary to find the *absolute slopes* of the isobaric surfaces. The main difficulty arises from the fact that the relative slopes or topographies of the isobaric surfaces are referred to the physical sea level whose topography is unknown and measurement is not yet feasible. However, it is possible from careful examination of the fields of temperature, salinity, and density to find a level where it can be assumed that the horizontal pressure gradients are either zero or small enough to be neglected. This assumed isobaric surface is selected as the *reference level* to which the slopes of the isobaric surfaces above and below it are referred. It is thought that the resultant dynamic force may resemble, to a certain degree, that which might be expected to exist in the area. This reference level is also known as the *level of no motion.*

However, the limitations of this approach have to be stressed in the sense that the barotropic mode of motion is not detected. The best way to calibrate the field of pressure is through direct current measurements along the water column. This method provides direct information on the absolute field of pressure existing at the time when the observations are made.

Current charts based on dynamic studies of the field of pressure have been used successfully for many years by the International Ice Patrol in the North Atlantic for anticipating the mean drift of icebergs. Other comparisons of the field of motion inferred from the relative field of pressure by means of the geostrophic approximation and from direct measurement of currents have shown that the differences may amount to from 5 to 25 percent with the mean departure about 15 percent.

But for all these discrepancies, it should be stressed that the method of the dynamic sections provides information on the *relative motion* to be expected in association with the baroclinic component of the field of pressure interpreted according to the assumptions of the geostrophic approximation. The barotropic component of the field of pressure can be obtained by comparing the computed geostrophic baroclinic flow with direct measurements of currents.

The normal procedure used to obtain the geostrophic field of motion from an oceanographic cruise in a particular area is carried out as follows: from the observed data at each oceanographic station, the oceanographic element values at standard depths are interpolated and the density (σt), anomaly of specific volume, and dynamic height (ΔD) are computed. A typical oceanographic station data report is shown in Table 4.1. Then convenient sections are selected and the vertical distribution of temperature, salinity, density, anomaly of specific

TABLE 4.1 Calculation of Anomalies of Specific Volume and Dynamic Depth*

Depth, m; pressure, decibar	Temperature, $t°$	Salinity, $‰$	Density, σ_t	$10^5 \Delta\sigma_t$	$10^5 \delta_{S,p}$	$10^5 \delta_{S,t}$	$10^5 \delta$	ΔD	$\sum_{po}^{ps} \Delta D$
0	22.11	36.40	25.27	271.3	0	0	271.3	0.0658	0
25	22.00	36.59	25.45	254.1	0.1	0.9	255.1	0.0618	0.0658
50	21.50	36.65	25.63	237.1	0.1	1.8	239.0	0.0579	0.1276
75	20.95	36.67	25.79	221.8	0.2	2.7	224.7	0.0547	0.1855
100	20.43	36.67	25.93	208.6	0.3	3.5	212.4	0.1031	0.2402
150	19.85	36.66	26.08	194.3	0.4	5.2	199.9	0.0973	0.3432
200	19.25	36.62	26.21	181.9	0.5	6.8	189.2	0.0922	0.4406
250	18.60	36.56	26.33	170.6	0.6	8.4	179.6	0.0885	0.5328
300	18.20	36.52	26.40	163.9	0.7	9.8	174.4	0.1722	0.6212
400	17.57	36.42	26.48	156.3	0.8	12.8	169.9	0.1662	0.7934
500	16.58	36.25	26.59	145.9	1.0	15.5	162.4	0.1527	0.9596
600	14.70	35.97	26.81	125.0	0.9	17.1	143.0	0.1336	1.1124
700	12.48	35.65	27.01	106.0	0.7	17.6	124.3	0.1160	1.2460
800	10.30	35.33	27.18	89.9	0.4	17.5	107.8	0.0976	1.3620
900	8.06	35.11	27.38	71.0	0.2	16.1	87.3	0.0756	1.4596
1000	5.90	35.02	27.60	50.1	0.0	13.8	63.9	0.1128	1.5352
1200	4.55	35.00	27.75	35.9	0.0	12.9	48.8	0.0956	1.6480
1400	4.09	34.98	27.78	33.1	0.0	13.8	46.9	0.0928	1.7436
1600	3.80	34.97	27.80	31.2	−0.1	14.7	45.8	0.0916	1.8364
1800	3.61	34.96	27.81	30.2	−0.1	15.7	45.8	0.0906	1.9280
2000	3.43	34.96	27.83	28.3	−0.1	16.6	44.8	0.1110	2.0186
2250	3.27	34.97	27.85	26.4	−0.1	17.7	44.0	0.1070	2.1296
2500	3.02	34.97	27.88	23.6	−0.1	18.1	41.6	0.2050	2.2366
3000	2.66	34.95	27.90	21.7	−0.2	18.9	40.4	0.1990	2.4416
3500	2.31	34.93	27.91	20.8	−0.3	18.8	39.3	0.2040	2.6406
4000	2.31	34.92	27.90	21.7	−0.4	21.1	42.4		2.8446

Atlantis station 1227; ϕ = 35°38′N, λ = 72°47′W, April 12, 1932. (From G. Dietrich and K. Kalle, *General Oceanography*, New York: Wiley-Interscience, 1953.)

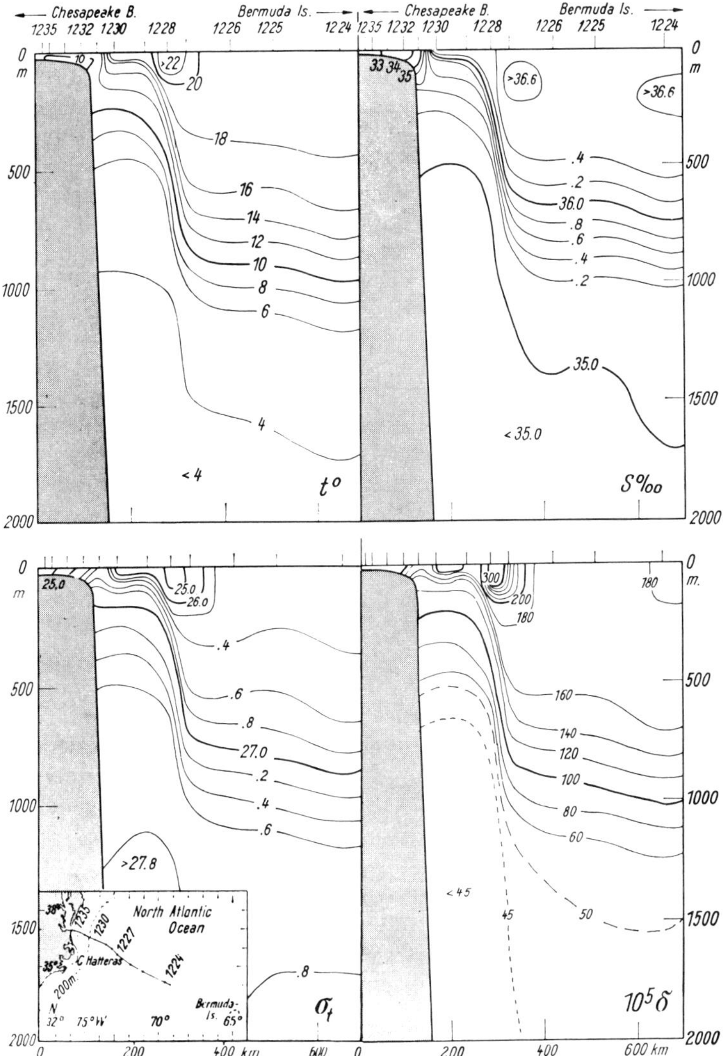

Fig. 4.1.   Distribution of temperature in °C, salinity, in parts per million, of density σ_t; and of the anomaly of specific volume in $10^5\delta$ along a section perpendicular to the Gulf Stream (see sketch of location) based on observations by the research vessel *Atlantis*, April 19–23, 1932 (vertical exaggeration 500-fold). (From G. Dietrich and K. Kalle, *General Oceanography*, New York: Wiley-Interscience, 1953.)

volume, and dynamic height of the isobaric surfaces are plotted. Some of these sections are shown in Fig. 4.1. Finally, the computed distribution of the geostrophic horizontal velocities and the associated volume transport are obtained. The geostrophic motion of the same section shown in Table 4.1 and Fig. 4.1 is shown in Fig. 4.2.

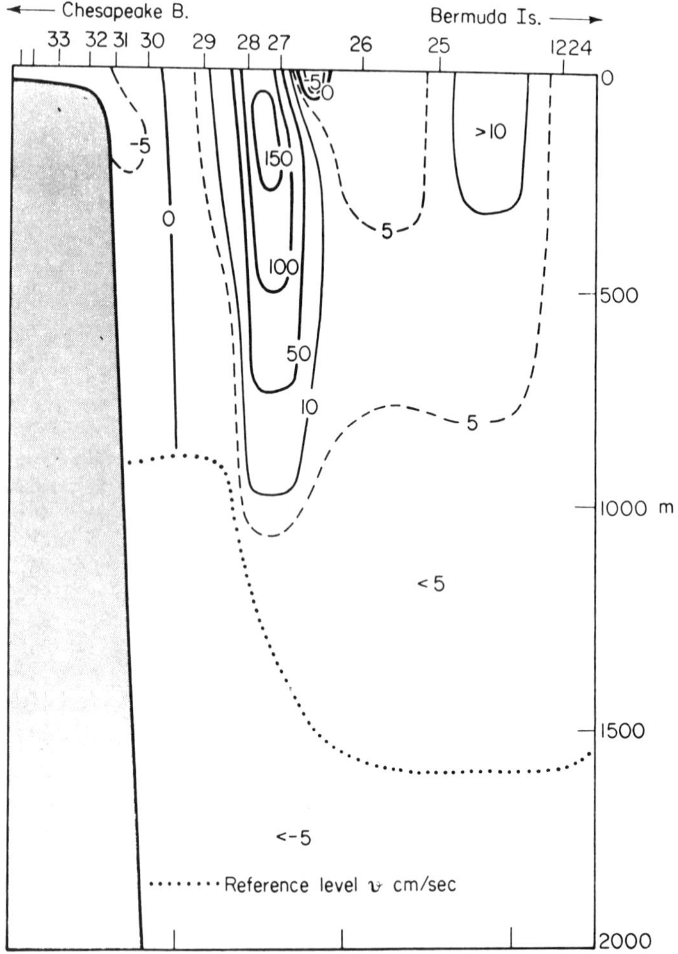

Fig. 4.2.   Current velocity (in centimeters per second) of the geostrophic current perpendicular to the section Chesapeake Bay–Bermuda Islands. No sign: northeast current, direction of Gulf Stream. Minus sign: southwest stream, countercurrent. (From G. Dietrich and K. Kalle, *General Oceanography*, New York: Wiley-Interscience, 1953.)

To obtain the same information for the whole area investigated, instead of particular sections, similar procedures are followed, although the data are now plotted on standard horizontal
charts of selected isobaric surfaces. The dynamic topography
of two selected isobaric surfaces from which the geostrophic
velocity is obtained are shown in Figs. 4.3 and 4.4; the area
covered in these two figures is the Gulf of Mexico.

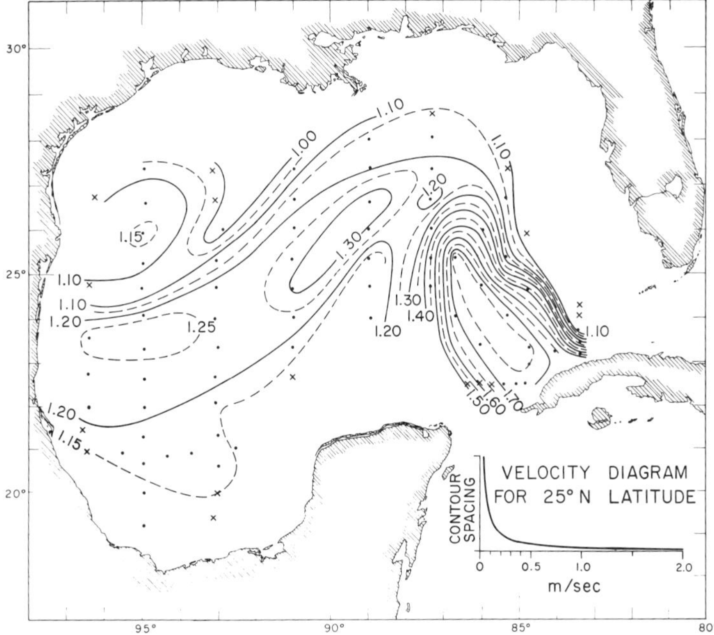

Fig. 4.3. Dynamic topography of surface relative to the 1000-
decibar surface, HIDALGO 62-H-3; X's indicate extrapolation. Contour interval, 0.05 dynamic meters. (From W. D. Nowlin and
H. J. McLellan, "A Characterization of the Gulf of Mexico Waters
in Winter," *J. Marine Res.*, Vol. 25, I, 1967.)

The Coriolis Force The generalized concepts of fluid
motion are based on the principles of Newtonian mechanics,
particularly the second law of motion. This equation treats the
motion of particles which can be essentially free of the earth—
or, expressed in another way, the particles of motion are referred to as an inertial coordinate system and are described in
their simplest terms. In general, particles of water or air move
with only very slight frictional influence from the rotating

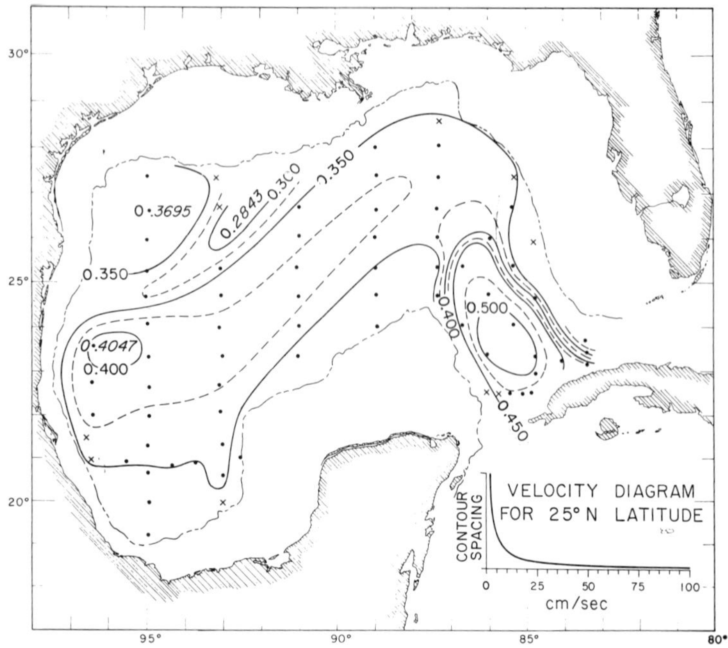

Fig. 4.4.   Dynamic topography of the 500-decibar surface relative to the 1000-decibar surface, HIDALGO 62-H-3; X's indicate some extrapolation. Contour interval, 0.025 dynamic meters. (From W. D. Nowlin and H. J. McLellan, "A Characterization of the Gulf of Mexico Waters in Winter," *J. Marine Res.*, Vol. 25, I, 1967.)

earth, and therefore they tend to follow paths which are most simply described by reference to an inertial system.

However, as the motion must be described for observers located on the rotating earth, it is more convenient to consider the motion relative to a rotating coordinate system. The effects of the earth's rotation are well known; in the northern hemisphere, the path of a body is deflected to the right of the observer and in the southern hemisphere to the left. The effect vanishes only when objects are moving horizontally and exactly eastward parallel to the equator. Ordinarily, the effects of the earth's rotation are not noticeable because the rate of the rotation is too slow in relation to motions observable in daily experience. However, in the case of winds and ocean currents, which move for hours and days without appreciable frictional influence from the solid earth, their paths must be strongly deflected. The effect may be so conspicuous that the paths of particle motion in the atmosphere and oceans may become highly circuitous or even successively looped in the course of several days. One way to account for this deflection is to consider the apparent curvature of motion of free particles as owing to a fictitious force acting at right angles to their direction of motion in the northern hemisphere. This fictitious force which performs no work is known as *Coriolis force.* The concept of a Coriolis force simplifies mathematical descriptions of nearly all the large-scale fluid phenomena of the earth, as well as the free motion of projectiles. The Coriolis force is a means of accounting for the behavior of particles moving in an inertial system but observed from a rotating earth which is assumed to be at rest.

The Coriolis acceleration is a function of latitude, and its magnitude is small, amounting to $1.5 \times 10^{-4}c$ at the poles and zero at the equator; here c is the horizontal velocity of a fluid parcel in any direction relative to the earth. For average winds, which have a speed of about 10 m per sec, the Coriolis acceleration amounts to 0.15 cm per sq sec or about 0.00015 g. In the case of strong ocean currents which move at about one tenth the speed of the average wind, that is, 1 m per sec or less, the Coriolis acceleration is one tenth as great. Small as this acceleration is relative to gravity, it has, nevertheless, important effects on the horizontal motions of the air and the water of the open sea. A principal counterbalance of the Coriolis force is the horizontal component of the pressure gradient force and the motion resulting from this balance is known as *geostrophic motion,* widely used in meteorology and oceanography.

The Coriolis acceleration is also responsible for another kind of motion observed quite often in nature and known as *inertial motion* or motion in an inertia circle. In small-scale processes which do not involve much change of latitude, inertial motion takes place in a circular path of a radius proportional to the velocity of the particle c and inversely proportional to the Coriolis parameter f. The period of such a circular motion is *half-pendulum day.* Motions in an inertia circle have been shown to occur where there is a sudden impulse which generates fluid motion and allows the system to coast without further interference.

The Coriolis acceleration, better identified by the Coriolis parameter f, plays an important role in other complex fluid phenomena associated with rotation, such as *planetary vorticity, inertia stability,* and *inertia waves.*

The Coriolis acceleration plays an important role in the basic hydrodynamic equations. When these equations are applied to large-scale phenomena, the variation of the Coriolis parameter with latitude is taken care of in the rectangular coordinate system by regarding it as a function of the horizontal coordinates. To simplify the equations, the Coriolis parameter is approximated by a constant except when differentiated with respect to a coordinate representing latitude, in which case the derivative is assumed to have a constant nonzero value, β. This simplification is known as β *approximation.* The β plane can be considered to represent midlatitudes of the earth. In equatorial and polar regions, the β-plane approximation is poor and other techniques should be used.

The β-plane approximation has proved to be a useful tool for dealing with some problems of dynamic oceanography, particularly in the time-dependent theory. However, it must be looked upon as an intermediate step in the development of more complete and satisfactory theories of ocean circulation.

FRICTION EFFECTS

The application of the hydrodynamic equations to oceanic currents requires the assumption of an ideal and frictionless fluid. Such fluids do not exist in nature, and where they rub against the solid earth or against each other, there is frictional coupling that cannot be ignored. Molecular friction is a process in which the momentum of rapidly moving particles of a fluid is exchanged with the momentum of relatively slower particles. The molecules are imagined to collide with one another to produce a diffusive exchange of energies from regions of rapid motion to regions of slower motion. This interchange of momentum produces a *shearing stress.* Stress, like pressure, is defined as force per unit area; but instead of being a force acting perpendicular to the area as it is in normal pressure, the stress acts parallel to the surface, i.e., tangential to the flow.

Whenever there is a variation of current speed in a direction at right angles to the flow, *velocity gradient*, an exchange of momentum takes place and shearing stresses are produced. The intensity of this exchange process depends on the velocity gradient or vertical shear and on the coefficient of viscosity. This coefficient provides an indication of the efficiency of exchange, since it determines how much stress will be produced by a given velocity profile. When the fluid is completely at rest or moving in a highly organized way (at very slow speeds), the rate of exchange or diffusion of heat, momentum, and dissolved solids is determined essentially by molecular motion. This condition is known as *laminar flow*. However, if this laminar flow is disrupted by increasing velocity or other stirring mechanisms, it is replaced by *turbulent flow* and the rates of exchange increase considerably. In the atmosphere and oceans, turbulent motion nearly always exists, and eddies in the flow can be so effective in moving particles among themselves that the effects of molecular diffusion are overwhelmed. For this reason, the *molecular coefficients* of viscosity, thermal conduction, and salt diffusion, which relate the gradients of momentum, heat, and salt concentration to the diffusive fluxes of these quantities as measured in the laboratory, are replaced by *eddy coefficients*, which are several hundred to many thousands times larger. Eddy viscosity is proportional to the velocity of the water within the eddies and to the size of the eddies. Characteristic values for the molecular and eddy coefficients for vertical transfer processes in the oceans are

Kinematic molecular viscosity ÷ 0.019 sq cm per sec
Kinematic eddy viscosity ÷ 2.0–7500 sq cm per sec

These figures quantify what has been said about processes in the ocean depending on the state of motion of its waters. The vertical eddy coefficients vary with the stratification of density and are greatly reduced where density increases considerably with depth.

Horizontal eddy coefficients vary over an even wider range from the order of 10 through 10^8 sq cm per sec. The explanation for this much more efficient exchange process may be attributed to the fact that horizontal motion can take place without appreciable work being done against gravity.

The distinction between vertical and lateral turbulence is particularly significant where the density of the water increases with depth (stable stratification). In this case, vertical random motion is impeded by Archimedean forces because a mass which is brought to a higher level will be surrounded by water of less density and will tend to sink back to the level from which it came and an opposite effect for water moving downward. Stable stratification reduces the vertical turbulence, and the eddy coefficients become smaller. The effect of stability on the lateral turbulence, on the other hand, is negligible because the lateral random motion takes place mainly along surfaces of equal density, and no work against gravity is performed.

BASIC HYDRODYNAMIC EQUATIONS

Classic hydrodynamics has been concerned with the motion of a homogeneous, incompressible fluid. When this body of knowledge is applied to real fluids such as air and sea water as they exist in nature, it becomes clear that because of many existent limitations it is difficult to explain their behavior. The main difficulty resides perhaps in the layered structure or stratification conditions present in those fluids under natural conditions. The lack of a systematic theory of the motion of a stably stratified fluid is one of the major obstacles to explaining meteorological and oceanographic phenomena by geophysical fluid dynamics.

The formulation of the basic dynamic equations to apply to the ocean involves the solution of the following problem: to find the relationship between acting forces and motion in a *stratified* fluid which is not a pure fluid but a *solution of salts in water* and to describe the motions in a *rotating system of reference*. Moreover, these equations will be used to describe motions present in the ocean circulation in different time and size scales.

In spite of the fact that this problem is attacked by making use of the classic tools of physics, in particular, dynamics and kinematics, it is desirable to clarify some of the concepts involved in the formulation of the various sets of hydrodynamic equations to solve the problem.

One of the concepts that needs to be discussed is how to specify the fluid motion. In oceanography, as in meteorology, the variables such as pressure, temperature, density, and velocity are functions of space and time. Usually, a fluid motion is specified in one of two ways, both equivalent: the *Eulerian description* or the *Lagrangian description.* In the first case, the fluid properties and the fluid motions are described at fixed points in the coordinate system. Therefore, the characteristics of the fluid to be described are only functions of time as their space coordinates are already specified. The set of specified points in space constitutes the *"network"* or grid of the property to be described, and the design of such a network depends on the size of the property to be studied. In the particular case of describing the motion, the set of isolines describing the velocity constitutes the *streamlines* of the motion.

The Lagrangian representation identifies the fluid elements by their position at some initial instant and the subsequent motion specified by the subsequent position and velocity of these fluid elements. The graphic representation of the motion in this case constitutes the *trajectories* of the fluid elements.

The selection of the system to be used is a matter of convenience. Many instruments, placed in points of the network of observing stations to measure fluid properties, provide Eulerian representation; while in questions related to diffusion or mass transport, the trajectory of the parcel is more adequate and a Lagrangian specification is more convenient. The systems are related, although for some particular problems, such as unsteady flow, the transformation has not been solved.

The motion of a fluid can be described by the conservation laws of momentum and mass, by the equation of state, and by the laws of thermodynamics. Moreover, these equations are of a very general nature and when applied to fluids such as the ocean and atmosphere, which may have some constraints or requirements for the particular phenomenon investigated, it is necessary to add the expression of such new requirements in the form of additional equations known as *boundary conditions.*

The first equation of motion is the conservation of momentum most commonly known as the second law of Newton. Every student of natural science is familiar with this law when applied to a solid object. But now it is applied to a continuous volume of fluid whose parcels may be moving at different velocities at different times. This must be accounted for and the equation for the conservation of mass meets this requirement. Some of the processes mentioned affect the density of the fluid, which modifies the pressure gradient force. This change in density of the chemical solution (sea water) produced by a change of any one of the three thermodynamic variables (pressure, temperature, and salinity) is taken care of by the equation of state and the laws of thermodynamics. Strictly speaking, the correct way to deal with the thermodynamic relationship for a sea-water system is the application of the first law of thermodynamics for a multiconstituent system within a gravitational field. The problem was greatly simplified with the introduction of the concept of salinity, which treats the dissolved salts as a single solute.

As for the forces which produce the motion, some are external and some pertain to the nature of the fluid itself. Force of gravity, the tide-producing forces caused by celestial bodies, and the stress of the wind on the water surface are external forces. Pressure gradient force, produced by changes in the density of the fluid, and the viscous stresses are typical internal forces. The important driving forces and the fictitious Coriolis force have already been discussed in more detail.

The concepts involved in the use of the hydrodynamic equations with geophysical phenomena have been discussed. However, the problem remains unsolved until the proper initial and boundary conditions are established. Some of the boundary conditions commonly used in fluid geophysical problems are that solid boundaries are impermeable or, in other words, the fluid does not flow through the walls of the container; this is the case of the sea water with sea bottom or the wind blowing through a mountain or out into space. The addition of fluid through some opening is a source, and the removal or flowing out is a sink. Also, in the case of the ocean and atmosphere, a layer separates them. The equations that describe the pertinent phenomena at the interface are to be formulated. In the case of viscous fluids such as air and sea water, the flow will finally cease unless energy is supplied from some external source to the system. It has been estimated that if solar radiation input were suppressed, all winds and ocean currents would practically cease in from 2 weeks to 3 yr. In this particular case, an energy equation describing this condition must be added to the set of equations.

Therefore, if the proper equations to describe the motion are selected, if mass is conserved, if all the forces relevant to the problem are considered, if sinks and sources of energy are included when necessary, and if the proper initial and boundary conditions are stated, it may now be possible to describe the motion of the particular geophysical fluid, provided that it is possible to solve the formulated set of equations.

In general, the complete set of hydrodynamic equations described cannot be solved exactly except in extremely simple cases. These equations are nonlinear, and the fundamental difficulty lies in the fact that the mathematical properties of nonlinear equations are not well understood. For this reason, oceanographers and meteorologists have frequently resorted to linearizing the equations by the assumption that the fluid motions are small perturbations on a steady flow. This is known as the *perturbation* method. When the motions cannot be considered as small perturbations, it is necessary to recur to numerical methods.

The methods used in oceanography to separate the steady-flow and time-dependent modes are almost analogous to those applied in turbulence theory. The principal difference is the consideration of the time and size scales of motion. In dealing with ocean circulation, the interest centers on the low-frequency, large-scale current systems for which the nonlinear interaction terms can be disregarded or simplified.

To have an appreciation of the scales of motion and the relative magnitude of the driving forces for a given flow, the steady-state equations are made nondimensional, introducing a characteristic length and depth, and some important coefficients are defined. The *Rossby number* is Ro, the *Reynolds number* is Re, and the *Froude number* is Fr. These coefficients and the slope of the ocean surface, s, evaluated for a given flow, provide an indication of the magnitude of the forces involved and facilitate the consideration of the important driving forces for a satisfactory description of the flow in terms of the equations. Another way to use these coefficients is to determine the horizontal and vertical scales for which any given term of the equations becomes comparable to unity. An example of how the Rossby number is applied to the ocean follows: existing data show that the maximum values of steady velocities are about 2 or 3 cm per sec; to make the Rossby number comparable to unity, it is necessary to assume that the current has a small characteristic length (sharply horizontally confined) or that it is flowing near the equator (Coriolis parameter, small). Concentrated western boundary currents, such as the Gulf Stream and Kuroshio, have a Rossby number comparable to unity as their horizontal scale of motion is of the order of 20 to 30 km for the maximum velocities indicated.

The other parameters, the Reynolds number and the Froude number, have values comparable to unity only at extremely small horizontal and vertical scales of motion.

The time-dependent equations can be studied in a similar way, although it is necessary to add a time scale to characterize the rate of variation and a density difference to show density variations. If the Rossby number of the steady flow is close to unity, the interaction terms representing convection by either the steady or time-dependent flow become comparable to pressure gradients and Coriolis forces. Therefore, in swift, concentrated currents, such as Gulf Stream and Kuroshio, the interactions between the two modes of motion cannot be ignored.

The basic hydrodynamic equations, where applied to a particular problem, are so rich in possible solutions and so complex that it is often difficult to decide how to apply them. This is when the oceanographer has to recur to his familiarity with ocean properties to properly formulate the problem. At the present time, various features of oceanic circulation are explained in different ways by changing the basic assumptions. These descriptions, which reach different conclusions, must be tested by observations in the ocean; and when reaching such a stage, attention should be given to making certain that the observational network is designed to give meaningful results of the problem under study.

STEADY-STATE CIRCULATION

The circulations of the geophysical fluids, particularly the oceans and atmosphere, are earth processes that have been continuously operating in a more or less smooth manner since the beginning of history. Therefore, it is not necessary to inquire how these motions began from a state of rest but to assume that steady-state conditions exist in which the flow at every point is not affected by time-dependent changes. This assumption is valid when the change that actually takes place over a short period of time is small compared with the accelerations required to initiate fluid motion from rest. It is convenient to investigate the steady solutions which are characterized by all time derivations to be zero, although any quantity such as velocity or pressure may vary in space. The question to be asked is the relevancy of such solutions, for the temporal changes observed in the atmosphere and oceans are striking and complex. Some of these changes, particularly the atmospheric ones, are overemphasized by the sensitivity of the human sensors, which may create a wrong impression about the adequacy of this approach. A 10 percent change in air temperature or pressure, which is a common fact in daily weather, will have very much larger effects on human metabolism. A less-biased consideration of these changes might rather stress the smallness of the perturbations from the steady-state condition. They are almost all below the 10 percent level.

In the ocean, the motion of the waters is characterized by fluctuations of a wide range of frequencies and size scales. The Gulf Stream, which has been repeatedly surveyed, is a typical example of a complex structure both in space and time. Deep water measurements also have shown a high degree of variability in the deep flow. Moreover, the processes taking place at the solid boundaries and at the air–sea interface are difficult to determine. To outline even the major dynamic relations that condition the behavior of an ocean current system, it is necessary to make drastic simplifications of the boundary conditions and of the exchange processes that drive the circulation.

Different models of steady-state motion have been developed in which the influence of friction, nonlinear acceleration, and thermohaline processes were considered separately to determine their effects. Attempts to combine them into one mathematical model have not been very satisfactory because of formidable analytical difficulties. It is concluded that steady circulation in the ocean has meaning only as an average with respect to time or space.

There are several kinds of motion that have been observed in large-scale circulation. The most prominent and representative of the atmosphere and the ocean are the geostrophic motions. The slow large-scale flow in the interior of oceans has traditionally been considered as steady and highly geostrophic, except in the upper 100 m where frictional stress from the atmosphere is significant. The geostrophic motion is generated when the fluid flow is both unaccelerated and frictionless. The only forces acting are the pressure gradient, Coriolis forces, and gravity. To be more precise, geostrophic motion occurs when the horizontal components of the pressure gradient and Coriolis forces are in approximate balance and frictional effects are confined to thin boundary layers.

The field of pressure in the ocean, as has been explained, is described by a set of horizontal charts showing the topography of the isobaric surfaces. The interpretation of the isobars in the pressure chart can be interpreted in the same manner as the topographic contours in a topographic map. Close spacing of the isobars indicates a stronger pressure gradient in the horizontal plane. It is necessary to set the particle in motion so that the balancing Coriolis force may begin to act and the geostrophic motion may be generated. To grasp the nature of this important motion in geophysics, it is convenient to follow the topographic analogy. Assume that a frictionless ball is standing on the top of a frictionless hill on the rotating earth (Fig. 4.5). Because of the component of gravity (pressure gradi-

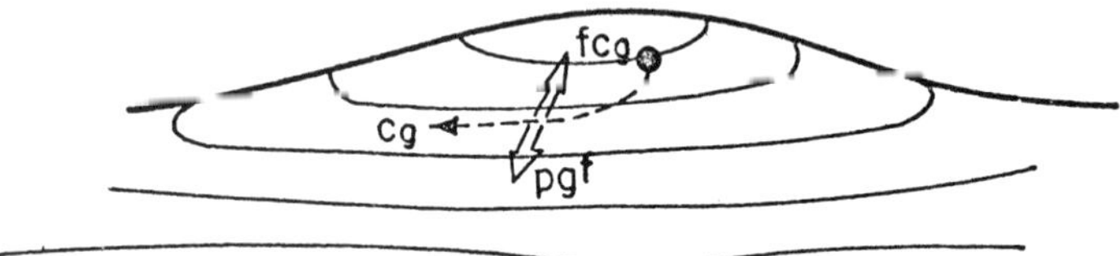

Fig. 4.5.   The geostrophic motion.

ent force), the ball slides freely across the topographic contours (isobars). As soon as the ball gathers speed, the Coriolis force acts and deflects the ball to the right (in the northern hemisphere). As the ball is deflected, the Coriolis force will always be perpendicular to the new deflected motion, and this will continue until the Coriolis force opposes and balances the downhill component of the gravity force. When this stage is reached, the ball will move along the topographic contours, and geostrophic motion has been generated. This geostrophic motion will continue indefinitely if friction is absent. But where friction acts, the ball speed is reduced and so will be the Coriolis force. Gravity will alter the course of the ball slightly downhill to gather the energy it needs to balance friction. The motion in this case will make an angle with the topgraphic contours toward the side of lower elevation (low pressure). In dealing with fluids, the geostrophic flow is along the isobars; and when friction is present, the flow is deflected and crosses the isobars toward the region of low pressure.

The geostrophic approximation is often used to compute the geostrophic field of motion from the scalar field of pressure. Given the set of charts showing the dynamic topography of the isobaric surfaces, the flow is along the isobaric contours with the highest topography to the right of the flow looking in the direction of the current (to the left in the southern hemisphere). The speed depends on the spacing of the isobars and is computed from the geostrophic formula. See Figs. 4.3 and 4.4.

Friction and turbulence exist to some degree in all motions of the atmosphere and the ocean. Horizontal accelerations are also present in any realistic analysis of these motions. The geostrophic balance must always be considered an approximation when applied to real geophysical phenomena. In some cases, as in the interior flow of the ocean, there is a good approximation; whereas in other cases, such as near the boundaries, it may require refinement before meaningful results are obtained. In other cases, it may be of no value whatsoever.

Another motion that has been observed to take place often in nature is the *inertial motion.* Some discussions about the nature of this motion were made when dealing with the Coriolis force. This motion occurs whenever a parcel of fluid moves in a curved path in such a way that the centrifugal force of the curved path is balanced by the centripetal acceleration due to the deflecting force of the earth's rotation. Inertial motion on the earth is necessarily anticyclonic. As the Coriolis parameter varies with latitude, there is a tendency to be more sharply curved in high latitudes than in low latitudes. This particularly produces a migration toward the west of the successive inertial circles. Figure 4.6 illustrates this type of motion.

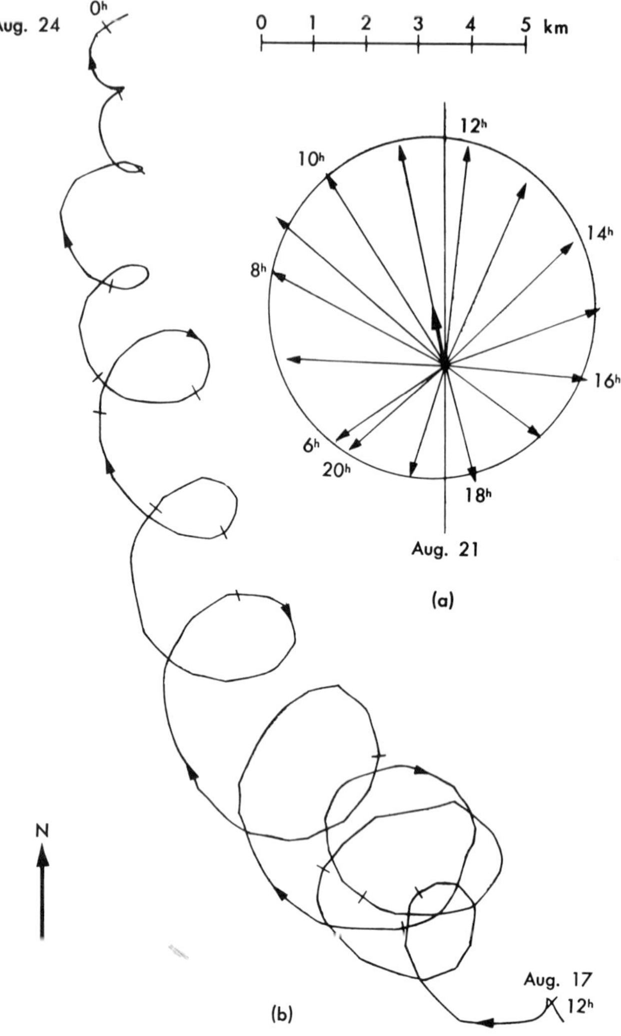

Fig. 4.6.   Rotating currents of period $\frac{1}{2}$ pendulum day observed in the Baltic, represented (a) by a progressive vector diagram for the period of 17–24 August 1933, and (b) by a central vector diagram for the period between 6 and 20 hr on August 21, according to Gustafson and Kullenberg. (From H. U. Sverdrup, M. W. Johnson, and R. H. Fleming, *The Oceans*, Englewood Cliffs, N.J.: Prentice-Hall, 1942.)

The steady-state circulation with consideration of friction is well represented by the *Ekman drift*. Friction, as exerted by the wind over the ocean surface, is the main driving force; and by viscous stresses within the subsurface water, the momentum is transmitted downward to a certain depth. The classic research of the wind-driven currents was done by V. W. Ekman, who, in 1905, in a theoretical paper, showed that the effect of the wind blowing steadily over an ocean of infinite depth and extent with constant vertical eddy viscosity is to drive the surface layer at an angle 45 deg to the right of the direction of the wind (to the left in the southern hemisphere). The water surface layer, in turn, by internal friction transmits the horizontal momentum to the layer immediately below, which is deflected to the right of the direction of the surface layer. This process goes on until a depth is reached where the direction of the water motion is opposite that at the surface. As the motion is transmitted downward and is deflected more and more to the right, the speed of the current decreases exponentially. The envelope of the projection of the velocity vectors in a horizontal plane is a spiral known as the *Ekman spiral*, as shown in Fig. 4.7. The depth where the direction of the flow is 180 deg from that at the surface is also considered the limit of frictional influence; the speed at this depth is about one twenty-third as great as that at the surface and a characteristic value is 100 m. The mean transport of water is at 90 deg to the right of the wind direction (to the left in the southern hemisphere).

The Ekman spiral has been shown to take place in the atmosphere and has been produced in the laboratory, but its occurrence in the ocean has not been conclusively demonstrated. The only observations available are on the displacements of fields of sea ice and recent field experiences with dye patches which move at some angle to the right of the wind.

Then there are some secondary effects associated with the Ekman transport, particularly when winds blow parallel or with a large component of their velocity parallel to the coastline. In the southern hemisphere, a northerly wind blowing along a west coast should drive the surface-layer water away from the coast. This water is replaced by colder water coming from the

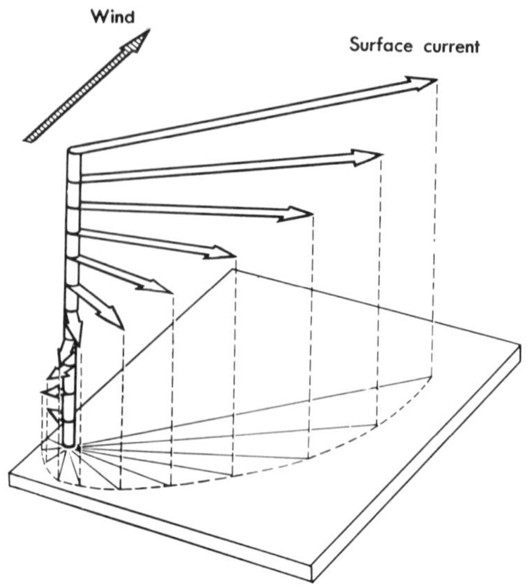

Fig. 4.7. Schematic representation of a wind-driven current in deep water, showing the decrease in velocity and change of direction at regular intervals of depth (the Ekman spiral). (From H. U. Sverdrup, M. W. Johnson, and R. H. Fleming, *The Oceans*, Englewood Cliffs, N.J.: Prentice-Hall, 1942.)

deep, and the process is known as *upwelling*. The opposite effect is accumulation and sinking of water along the coast. The biological implications of upwelling are of great interest because of increasing productivity through an increased supply of nutrients from below the euphotic level. Sinking of water along the coast also has biological and atmospheric implications. A dramatic example of this phenomenon takes place off the coast of Peru, where during February and March of certain years a southerly flow of air produces coastal sinking which permits the warm equatorial water to intrude along the coast. This current, known as El Niño, not only reduces considerably the yield of the fishery but may produce very undesirable climatic effects.

Besides the biological and other implications of these vertical motions of water, there is also a dynamic one which is commonly referred to as secondary effect of the wind. This effect is the generation of a surface current parallel to the wind direction by the deformation of the field of density produced by the upwelling or sinking of water.

The present state of the knowledge on wind drift can be summarized as follows:

1. The deflection angle at the surface is much smaller than the theory suggests. It is about 10 deg. What seems fairly certain is that the average flow in the Ekman layer must be 90 deg to the right of the wind (in the northern hemisphere) and that there must be some type of spiral in the current directions.

2. It is also clear that the bottom of the Ekman layer lies 100 m or so deep within a factor of 2 or 3.

3. Very little is known about the structure of the spiral and of the turbulent mechanisms that determine its nature.

TIME-DEPENDENT MOTION

It has been said that the ocean exhibits a degree of variability higher than suspected. Although there is a growing concern within the field of oceanography about ocean variability, the present knowledge of the time-dependent modes of response of the ocean is far from adequate. Most of the research on the time-dependent motion in the sea has been restricted to the study of the different types of waves that can be propagated in a homogeneous or two-layer ocean without interacting with the steady flow. Moreover, the interest is centered in wave modes with periods greater than a half-pendulum day. Waves of higher frequency have been studied more thoroughly in connection with tides and surface waves and swell.

The variations of the Coriolis parameter with latitude is responsible for the ocean developing a low-frequency wave motion that propagates toward the west relative to the particle motion. These waves, known as Rossby or planetary waves, provide the mechanism for moving energy on a time scale greater than a half-pendulum day; and because of some particular characteristics of the horizontal velocities in the waves, they can be considered as moving current systems. These low-frequency waves are instrumental in the description of ocean circulation as a steady circulation and in the transient response of the oceans to changes in the driving forces that maintain the steady circulation. However, the interactions between the steady state and the wave motion are little known. This is particularly important in the western boundary of the oceans, where the steady state is characterized by the presence of the fast and narrow western boundary currents. Here, the Rossby number is close to unity, and the degree of interaction is considerable; therefore an appreciable exchange of energy between the two modes may take place.

Studies of the forced response of one ocean of infinite extent and depth provide a qualitative description of the processes generated by variations of wind stress. However, it is thought that the results will be affected considerably by the addition of boundaries and components of the steady flow. An increased effort on the analysis of the time-dependent equations is necessary to better understand the nature of Rossby waves and the response characteristics of a partially or fully bounded ocean.

WIND-DRIVEN CIRCULATION

It has been well established that if a wind blows in a steady direction during a reasonable time, it is capable, through frictional stresses, to transmit momentum to water. The planetary wind field shows that the seasonal variations of the air circulation around the semipermanent centers of high and low pressure over the oceans are of considerable magnitude. The response of the oceans to these changes can take place in two different modes, the barotropic mode or the baroclinic mode. The nature of these changes was discussed in Chapter 3.

In spite of the fact that the ocean attempts to respond immediately to the variations of atmospheric pressure and wind stress by altering the barotropic circulation, it does not come to equilibrium with the seasonal wind field before changing seasons because of the slow baroclinic processes in the reorganization of the temperature and salinity fields. Because of this circumstance, the oceans are constantly trying to get in phase with the winds, and the amplitudes of seasonal fluctuations in ocean circulation are of smaller amplitude than the variations in the atmospheric circulation.

It is a well-proved fact that the general surface circulation of the ocean is almost coincident with the planetary wind field. This strong resemblance suggests a high degree of coupling between both circulations, or stated in other words, that the surface currents of the ocean are wind-driven. Since the classic work by Ekman in 1905 on the wind drift, several investigators advanced theories describing different aspects of the wind-driven circulation. The most attractive one is the vorticity theory which is able to prove that the torque of the planetary field on the sea surface generates a field of similar motion in the ocean volume.

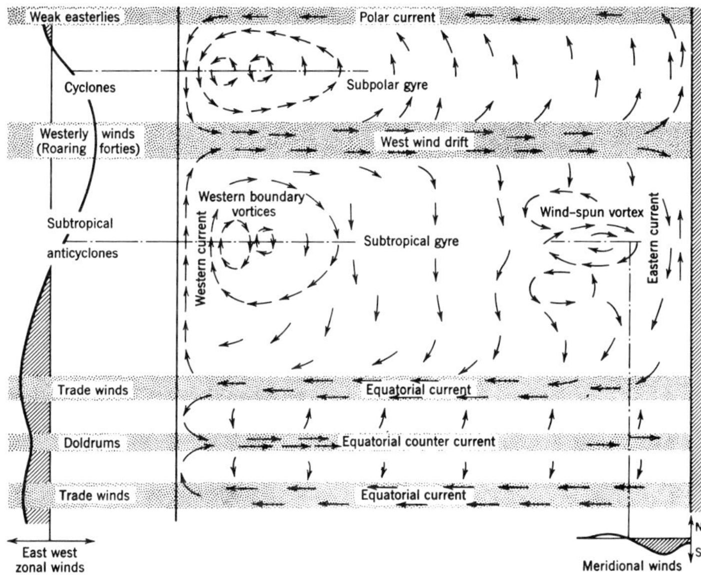

Fig. 4.8. Schematic presentation of circulation in a rectangular ocean resulting from zonal winds, meridional winds, or both. The nomenclature applies to either hemisphere, but in the southern hemisphere the subpolar gyre is replaced largely by the Antarctic Circumpolar Current (west-wind drift) flowing around the world. (Reprinted from W. H. Munk, *J. Meteor.*, Vol. 7, 89, 1950.)

Munk (1950) computed the pattern of currents that the earth wind systems would generate using the vorticity model. Figure 4.8 describes the major features of ocean circulation as calculated by Munk. The strong western boundary currents correspond to the Gulf Stream and Kuroshio in the North Atlantic and Pacific, respectively, the Brazil current in the South Atlantic, the Mozambique and Agulhas currents in the Indian Ocean, and the East Australia current in the South Pacific. In the Southern Ocean, the *west-wind drift* generated by the "roaring forties" flows without interruption around Antarctica and corresponds to the Antarctic circumpolar current. The equatorial system deserves a more detailed explanation because at the equator the horizontal component of the Coriolis force vanishes and the sense of the deflection reverses, crossing from one hemisphere to the other, and its magnitude increases rapidly with increasing latitude. These peculiarities generate two regimes of ocean circulation in the trade-winds region. In the equatorial calms or doldrums, between the two trade-wind belts, the predominant wind is from the east but of much less intensity than the trades. The water driven to the west by the trade winds piles up on the east coast of the continents. This westward flow forms the equatorial currents. The water that has accumulated against the eastern border of the continents flows poleward into both hemispheres, but some of this water flows back (eastward) in the doldrums region, forming the equatorial countercurrents. Increased efforts in the equatorial regions resulted in the discovery of strong eastward flow in the subsurface layers at the equator and at a mean depth of 100 m. These are known as the *undercurrents* of the equatorial system. The Cromwell Undercurrent in the equatorial Pacific transports around 40 million cu m of water per sec, which ranks it among the major currents of the world ocean.

An appropriate way of demonstrating the importance of a current system in the geophysical world is by the rate of water volume being transported. This moving water carries with it its proper concentrations of salts, heat, and momentum to other regions of the world ocean. Figure 4.9 shows the transport of water above the main thermocline in the North Atlantic in which some of the characteristics of the circulation discussed can be seen.

THERMOHALINE CIRCULATION

In spite of the successful approach in describing the broad picture of the surface oceanic circulation and some of its particular features by the wind-driven vorticity model, some difficulties remain. The modern conception of the convective mode of circulation can account for the inconsistencies of the

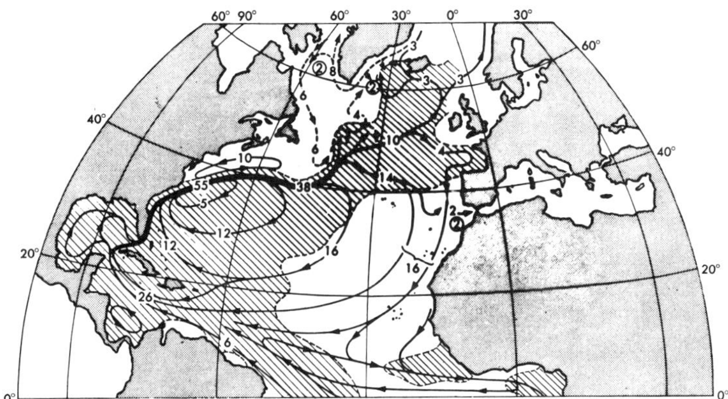

Fig. 4.9. Transports of water above the main thermocline in the Atlantic Ocean. The lines with arrows indicate the direction of the transport. The inserted numbers indicate the transported volumes in millions of cubic meters per second. Dashed lines show cold currents; continuous lines show warm currents. Areas of positive surface temperature anomaly are shaded. (From H. U. Sverdrup, M. W. Johnson, and R. H. Fleming, *The Oceans,* Englewood Cliffs, N.J.: Prentice-Hall, 1942.)

advective mode and complete the picture of the general oceanic circulation.

The mechanism of convective circulation in the world ocean consists of the following processes:

1. The surface water is continuously interacting with the atmosphere and exchange of energy and matter takes place across the boundaries. The variability of the characteristics of both fluids prevents the establishment of a complete thermodynamic equilibrium.

2. The surface water is conditioned by the different exchange processes, and when considered in large-time scale, oceanic climatic regions can be distinguished. Those processes which take place across the air–sea interface and which affect the surface water characteristics were discussed in Chapter 3.

3. The conditioning of the surface water is such that at high latitudes the surface water tends to sink as it is cooled and to slide toward the equator at depths in agreement with its density. They are the sources of deep water masses.

4. The equatorward circulation of deep water takes place mostly through narrow deep western boundary currents (along the eastern side of the continents).

5. Slow rising motions required to maintain the main thermocline (distributed sink) close to the convective cycle.

The surface waters of the world ocean can be divided into three great water masses: the equatorial water of the midtropical region, the central water found in middle latitudes, and the subpolar water of the high latitudes.

The processes at the surface that give rise to the water masses listed and the sinking under action of climatic stresses are considered a continuous or repetitive activity which may be responsible for large masses of those water types. If the water masses in the depths of the ocean were not renewed by water that was already in contact with the atmosphere and saturated with oxygen, the concentration of this element would eventually reduce to zero and the other properties and concentrations would diffuse through vertical mixing with the other layers; in the course of time all those properties and concentrations would be uniformly distributed. The fact that the water shows a constant of composition and a vertical gradient of salt concentration in sea water speaks eloquently of the effectiveness of horizontal mixing processes and of the resistance of the water of the present oceans to vertical exchanges.

The main driving force in this convective motion seems to result from the balance of the thermal and evaporation–precipitation processes. The available data do not permit us to determine whether the thermal or haline mode of increasing the surface density predominates; the whole process is called the *thermohaline circulation*. The average conditions of the thermohaline circulation in the oceans are schematically shown in Fig. 4.10.

Stommel (1957) suggested a hypothetical circulation in the deep water of the world ocean which is also compatible with the wind-driven circulation above the main thermocline. He found two principal sources of deep water, one in the area of the Labrador and Irminger seas and the other in the Weddell Sea in Antarctica.

The water that sinks in the Labrador area is being constrained into a narrow deep current flowing along the east coast

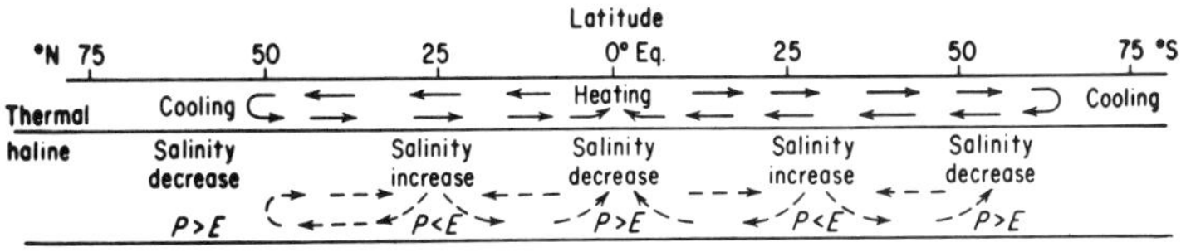

Fig. 4.10. Representation of average sources, and sinks, for a thermohaline circulation in the ocean. For latitudes poleward of 25 deg, the effect of salinity changes on density counteract the effects of heating and cooling. An efficient thermohaline circulation seems to be restricted to the upper layers in tropical and subtropical regions. (From G. Neuman and W. J. Pierson, Jr., *Principles of Physical Oceanography*, Englewood Cliff, N.J.: Prentice-Hall, 1966.)

of North and South America to the latitude of the subtropical convergence where it may turn toward the east, cross the Indian Ocean, and mix with the Pacific deep water mass. The water that sinks in the Weddell Sea, moves northward as a western boundary current along the eastern coast of South America up to 40°S latitude where it joins the southbound flow. Here, Antarctic intermediate water mixes with South Atlantic deep water and, turning eastward, follows the route mentioned.

The thermohaline model of abyssal circulation gives strong emphasis to the nature of the circumpolar current which flows

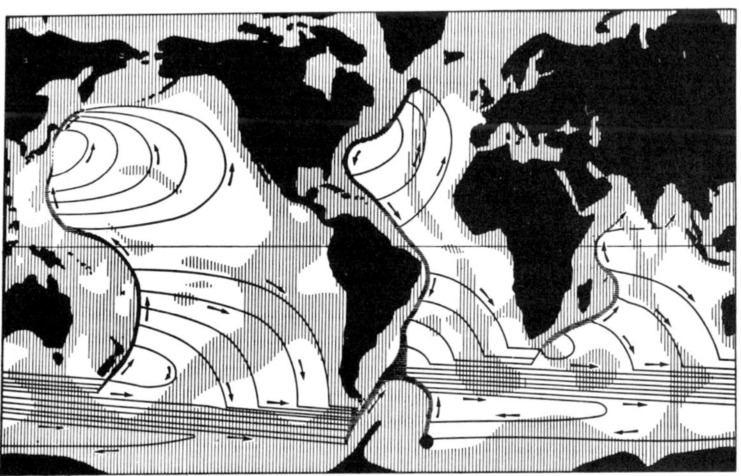

Fig. 4.11. The deep circulation of the world ocean according to Stommel. (From H. Stommel, *Deep Sea Res.*, Vol. 5, 82, 1957).

with practically no obstruction around Antarctica and which
Stommel considers instrumental in the exchange between the
major ocean basins. The water transport has been computed
to be around 120 million cu m per sec, which is the greatest
transport taking place in any current on earth.

As a result of the latest developments in theory, observations,
and experiments, it is thought that the general oceanic circulation consists of a wind-driven gyre under each pair of the major
zonal wind belts and a planetary pattern of thermally generated
deep transport system that is consistent with the wind-driven
circulation above the main thermocline. Figures 4.11 and 4.12
illustrate the abyssal oceanic circulation and its relation to the
wind-driven surface circulation.

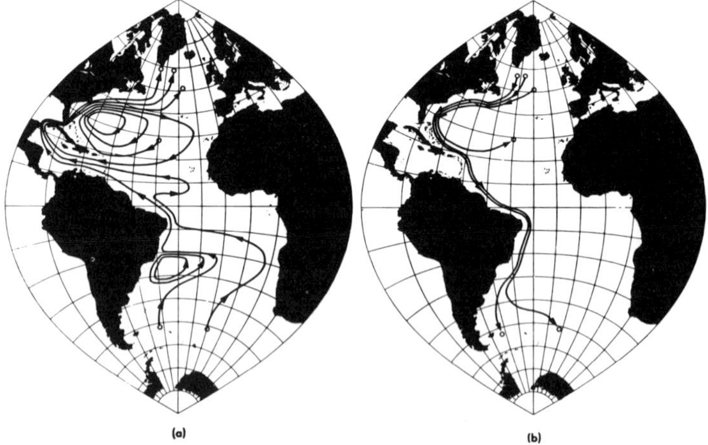

Fig. 4.12.   Relationship of the surface (a) and deep (b) circulations
of the Atlantic Ocean according to Stommel. (From H. Stommel,
Deep Sea Res., Vol. 4, 161, 1957.)

SUMMARY

The large-scale currents of the ocean, where considerable
transport of water is involved, can be accounted for by two types
of circulation. The advective mode, or wind-driven circulation,
and the convective type, or thermohaline circulation. The pattern of the first type is very similar to the pattern of the surface
winds, which are the main driving force. Besides this striking

resemblance, the wind-driven circulation is characterized by narrow, swift, jet-type currents along the western side of the oceans. They are known as *western boundary currents*, and the Gulf Stream, the Brazil, and Kuroshio currents are typical examples of this type of current. The earth's rotation is responsible for this feature. The convective mode, or thermohaline circulation, closes the circuit and is explained as the result of water being conditioned at the surface by climatic factors which sinks and spreads at different depths. Thermal effects on the surface waters give rise to the sources of dense water which will reach the depths of the ocean. This type of circulation is characterized by a very broad, slow flow with a diffusion of matter upward to compensate for the sinking. As in the case of wind-driven circulation, a stronger deep flow takes place at the western boundary of the ocean. The earth's rotation is also responsible for this particular feature.

Because of the increasing interest in long-range weather forecasting, the interaction between the ocean and the atmosphere has been given considerable attention. The understanding of how the ocean reacts to the different meteorological and even climatic conditions had increased particularly in the Ekman layer. For the activities that the oceanographic engineer has to carry on in the ocean, the following environmental characteristics, related to ocean circulation, should be stressed:

1. The region of strong and variable flow is the Ekman layer, or wind-drift layer. All the processes occurring at the air–sea interface are felt in this layer. The current intensity is greatest at the surface and decreases until it practically ceases at the bottom of the Ekman layer.

2. The tidal currents, which are of much lesser importance insofar as volume transport is concerned, are restricted to shallow waters where they may reach very high values, up to 15 knots. They can be predicted with very good accuracy, and the flow intensity is similar from the surface down close to the bottom.

3. Other important currents associated with periodic phenomena, such as surface waves breaking on the beach (longshore currents) and currents associated with internal waves, are discussed in Chapter 5.

5

The Waves of the Sea

In Chapter 4, the general oceanic circulation was described. The steady flows of both the advective and convective circulations carry with them impressive amounts of matter, energy, and other concentrations, and in the environmental implications of these transports reside the great interest in understanding general oceanic circulation. There are other modes of motion in the sea which are characterized by some periodicity and are commonly known as waves. The most familiar are the surface wind waves observable over the sea. What is typical of wave motion is that waves transport energy from one place to another without transport of matter. The energy contained in the water coming from a faucet is clearly discernible as the flow of water is capable of moving objects when it strikes them. However, a boat floating on the surface of the water will move up and down when subjected to wave action but not be displaced bodily with the moving deformation of the water. The energy capable of moving the boat up and down reaches it without a steady flow of matter. In a similar way, the waves generated by a storm in the North Atlantic will destroy some beaches in England a few days later even though no water has been displaced from the generating area to England. What impresses one most concerning surface waves is their rapid motion across the surface, which creates a wrong impression that the water itself is moving forward at such a speed. If this were the case, no structure could be built to be strong enough to stand such a pounding. Water moving at high speed feels almost as solid as concrete.

Wave motion, then, is a propagation of a disturbance with no propagation of matter and can be generated in any medium capable of reacting to any deformation with a restoring force that tends to suppress the deformation. The surface of the sea lends itself to the most obvious forms of wave motion because it resists being crested up or hollowed out in a heap. The restoring forces are either gravitational or due to the effect of surface tension, and the nature of the resulting motion depends on the disturbing force and on which of the two restoring forces opposes the disturbing one. If gravity prevails, the generated waves are called gravity waves. If surface tension is the restoring force, the waves are capillary waves.

A very convenient way to see what really happens in a wavy motion is by watching a bottle floating low in the water. The combined motion of moving up and forward with the passing of the crest and down and backward in the trough is very close to a circle. The difference in elevation between the wave crest and the wave trough is called the *wave height*, H, and the time required by the bottle to return to its almost original position is the *wave period*, T. The distance from crest to crest (or trough to trough) is the *wavelength*, L. The phase speed c (or time required by one crest to move one wavelength is $c = L/T$. The frequency $f = 1/T$ (in cycles per second) is commonly used. The orbital velocity or water velocity cp is the length of the circumference of diameter H divided by the period T. The relation between the water velocity and the phase velocity is $cp/c \approx 3H/L$. There is not necessarily a connection between the period and height of the wave. Wave heights are much smaller than wavelengths, and the stability of the waves is set by the relation wave height–wavelength (H/L). Stokes' theory put an upper limit value $H/L = \frac{1}{7}$ for the crest to become unstable and start to break. The water velocity beneath the crest is in the same direction of the phase velocity, while under the trough it is in the opposite direction. The forces related to these alternating motions are small compared to other forces. Below the surface in deep water, the particle motion is rapidly attenuated. At a depth equal to half the wavelength, the particle motion is $\frac{1}{23}$ its surface value; at a depth of one wavelength, it is $\frac{1}{530}$ the magnitude at the surface. The information about depth of wave action is of particular importance to geologists and engineers. What may look like a small velocity for some particular purpose may be strong enough to move sediments and rework bottom topography. Theoretical calculations show that a 3-m (10-ft) wave, 150 m (500 ft) long, would move sand in 65 m (192 ft) of water.

Gravity waves are normally dispersive; that is, their velocity of propagation or celerity of phase propagation depends on the wavelength. For sinusoidal waves, the wave equation shows that the phase velocity is a function of root square of the *relative depth* h/L, h being the depth of the water. On the basis of this relative depth, waves are classified as *deep water waves* when $h/L > \frac{1}{2}$ —the depth of the water is larger than half the wavelength—and *shallow water waves* when $h/L < \frac{1}{20}$. For deep water waves, the general formula for the celerity of phase propagation reduces to $c^2 = gL/2\pi$; and for shallow water waves, $c^2 = gh$. So, from the above formulas, it can be seen that deep water waves travel faster as their wavelength increases; they constitute a dispersive system. Shallow water waves travel with the same speed whatever their wavelength may be. Figure 5.1 shows the variation of the phase speed with wavelength.

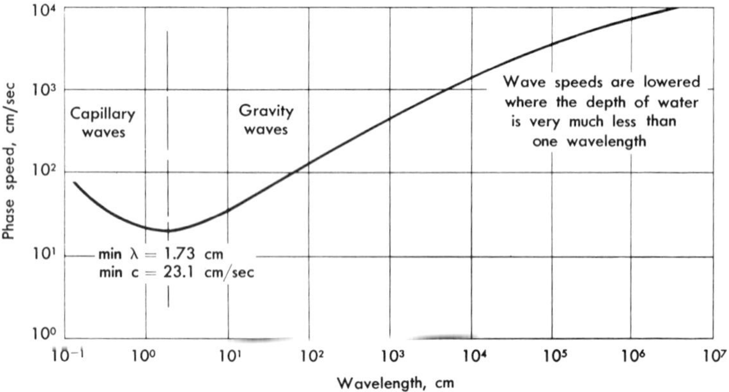

Fig. 5.1.   Celerity of gravity waves and capillary waves as a function of their length. (From W. S. Von Arx *Introduction to Physical Oceanography*, Reading, Mass.: Addison-Wesley, 1962.)

The characteristics of waves, as they occur in nature, are very difficult to represent in simple terms. This is the reason they are described by the interactions of minute, long-crested, sinusoidal undulations on the surface of a perfect liquid which occur in indefinitely long trains of equal amplitude that move without loss of energy over an infinite distance for an indefinite time. The simplest way to interact is the case of two identical wave trains moving in exactly opposite directions; standing waves are generated which have the same period and wavelength as the original trains but a greater range of amplitudes. If such a train moves with a slight difference in direction, the height of the resultant wave train is increased and the waves become *short-crested*. The next step, to go closer to natural conditions, is to consider two trains of sinusoidal waves of the same amplitude but very slight difference of frequency moving in the same direction. The resultant train is characterized by a succession of beats as the phases of the two *long-crested* wave trains interact to change the local amplitudes. Such beating produces a succession of wave groups uniformly spaced and separated by bands of almost undisturbed water. The complexity of the wave spectrum at sea is better demonstrated when the two wave trains of almost identical frequency are moving in slightly different directions. The complexity of this natural phenomenon makes it difficult to properly use some of the wave concepts stated before, such as wavelength, wave direction, frequency or period, and phase velocity. However, it has proved very fruitful to study the properties of natural gravity waves in terms of the ideal waves, namely, these terms are a train of long-crested sinusoidal or trochoidal waves of small amplitude moving on a perfect fluid of uniform depth, h.

A concept which requires some amplification is the one related to wave speeds. Typical of a system of waves of different frequencies propagating in a dispersive medium or a dispersive wave system moving in any kind of medium is, as often stated, the concept of *group velocity*. Perhaps the best way to visualize the problem of the wave speed is to produce a disturbance by an impact and study the wave system that will propagate from the initially concentrated impulse applied to the free surface. This is known as the *Cauchy–Poisson problem*. The situation is similar to dropping a rock into still water. The energy put into the water by the impact is set in motion and is stored as potential energy in the deformation of the water surface and as kinetic energy in the motion of the deformation. The generated wave train is composed of shorter waves in the rear and longer at the front. If one tries to follow by visual observation an individual wave crest—the last one in the train—it is soon found that another wave was created behind it and it now is next to last. A bit later, the wave being followed will be the third from last, and so on. As new waves are growing up at the rear of the train, wave crests are vanishing at the front. This will be the fate of the wave being watched. Therefore, it is concluded that the group of waves moves in two ways. The individual crests move forward and gradually disappear and the group propagates by adding crests at the rear and subtracting at the front. The group remains as an object that can be identified and which can have a meaningful speed but with constant change in the individual waves. The result is to move the group more slowly than the individual crests. This velocity is called the *group velocity*.

This behavior of a wave group moving with group velocity has also been found in other mathematical models such as the *Gaussian wave* packet and the *Pierson finite wave train*, in a tridimensional space. The physical interpretation of the group velocity can be related to the transport of energy. It can be demonstrated that if waves of constant height are generated, the wave front which advances with phase velocity carries a small part of the energy. The energy, in appreciable amounts, moves at the group velocity. For deep water waves, the group velocity is half the phase velocity; and for shallow water waves, the group velocity is similar to the phase velocity. Another interpretation of the group velocity evolves from the linearized approach to the wave problem. It can be proved that if two interacting waves become more and more nearly alike in both length and frequency, the speed of the amplitude-modulated wave approaches the group velocity.

There are also two well-known modes of oscillation in the classic wave theory that occur often in the ocean, particularly in natural embayments, canals, harbors, and ports. They are represented by the *standing wave* and the *progressive wave* types. Their orbital motion, streamlines, velocity distribution for deep, intermediate, and shallow water waves are shown in Figs. 5.2 and 5.3.

OCEAN WAVE CLASSIFICATIONS

Ocean surface waves are classified in a number of useful ways. The most common is by period or frequency. Another way is

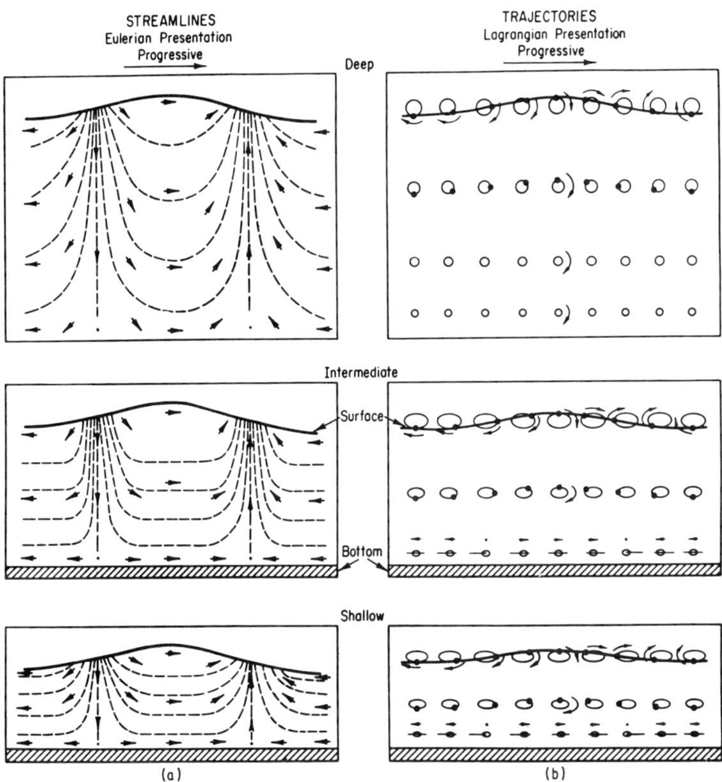

Fig. 5.2. Particle paths (Lagrangian viewpoint) and streamlines (Eulerian viewpoint) for a sinusoidal progressive wave. (From B. Kinsman, *Wind Waves*, Englewood Cliffs, N.J.: Prentice-Hall, 1965.)

by the generating force, such as wind waves, tsunamis, astronomical bodies, and so forth. However, more important than the disturbing force is the degree of interaction between the applied force and the mode of oscillation. In this regard, the waves can be classified as *free waves* or *forced waves*. A free wave is generated by the sudden application of the disturbing force which is removed immediately after producing the disturbance. The mode of oscillation will depend on the characteristics of the oscillating system. The waves generated by throwing a stone in the still water belong to this type. In a forced wave, the generating force is applied continuously and the system will be forced to oscillate tuned to the force with modifica-

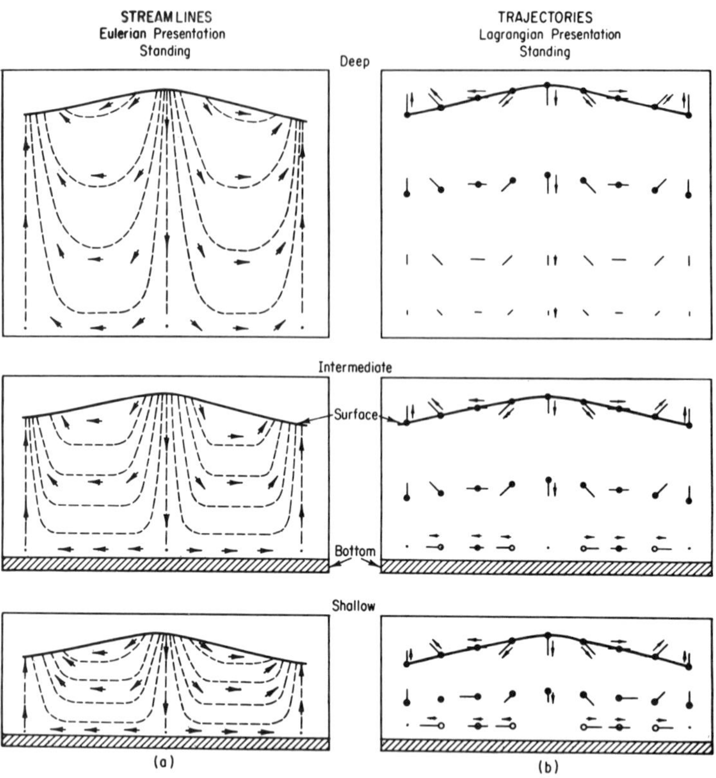

Fig. 5.3. Particle paths (Lagrangian viewpoint) and streamlines (Eulerian viewpoint) for a sinusoidal standing wave. (From B. Kinsman, *Wind Waves*, Englewood Cliffs, N.J.: Prentice-Hall, 1965.)

tions imposed by the oscillating system. The classic example of a forced wave is the tide. Its period is conditioned by the periods of the moon and sun and by the inertia of the water, resonant characteristics of the excited body of water among other restraints. Wind-generated gravity waves are not as clear-cut a case as the tides. In the stormy area, the generating force is considered as a sudden impulse applied to the water. However, as they move, the wind is accompanying them, so it cannot be considered a free wave because the wind is still working on the waves. This interaction between free and forced waves is very complicated. When the wind waves leave the generating area and propagate like a swell, they behave more like a free wave.

Another way of classifying waves is by the other force present in any oscillation, that is, the restoring force. The main restoring forces for the ocean waves are surface tension, gravity, and Coriolis.

It has been shown that the motion of the wave profile is different from the motion of the water particles. The traditional classifications of *longitudinal* and *transversal* waves in physics do not apply in ocean waves. They belong to neither class; the particle motion is circular in the vertical plane. The deep water wave or short wave and the shallow water wave or long wave based on the relative depth h/L as well as the progressive and standing waves have been discussed. Figure 5.4 shows the waves divided in some of the ways indicated above with the addition of a curve showing the energy that may be contained in the different frequencies (power spectrum).

The shortest-period waves in the spectrum are the capillary waves. They are the first to be observed when the wind starts to blow on the water surface or on the backs of ordinary gravity waves. They look like a fine structure of small ripples of nearly capillary dimension. It is thought that these waves considerably contribute to the energy transfer from the wind to the water, since they are very numerous and move slowly before the wind. They also play an important role in the reflection, refraction, and scattering of sound and light. Their wavelength is shorter than 1.73 cm and the phase velocity decreases with increasing wavelength, which is the case of *anomalous dispersion* (Fig. 5.2). The minimum celerity is 23 cm per sec, which corresponds to a wavelength of 1.73 cm. The restoring force is surface tension and the surface of the water behaves like a stretched membrane. Very short waves have much more curvature per unit of length than longer waves, and this explains why surface tension is instrumental in resisting the deformation. The *sea clutter* observed in radar return is more closely associated with the smaller ripples and capillary waves than with longer wind waves. The implications of capillary waves in radar, in glitter measurements, and in air–sea interface, gas–salt exchange gives evidence of the growing interest in this region of the spectrum.

The ordinary gravity waves observed at sea, which cause the rocking of a ship and break along the coasts of the continents, are composed of two types of waves. The first are *sea* waves,

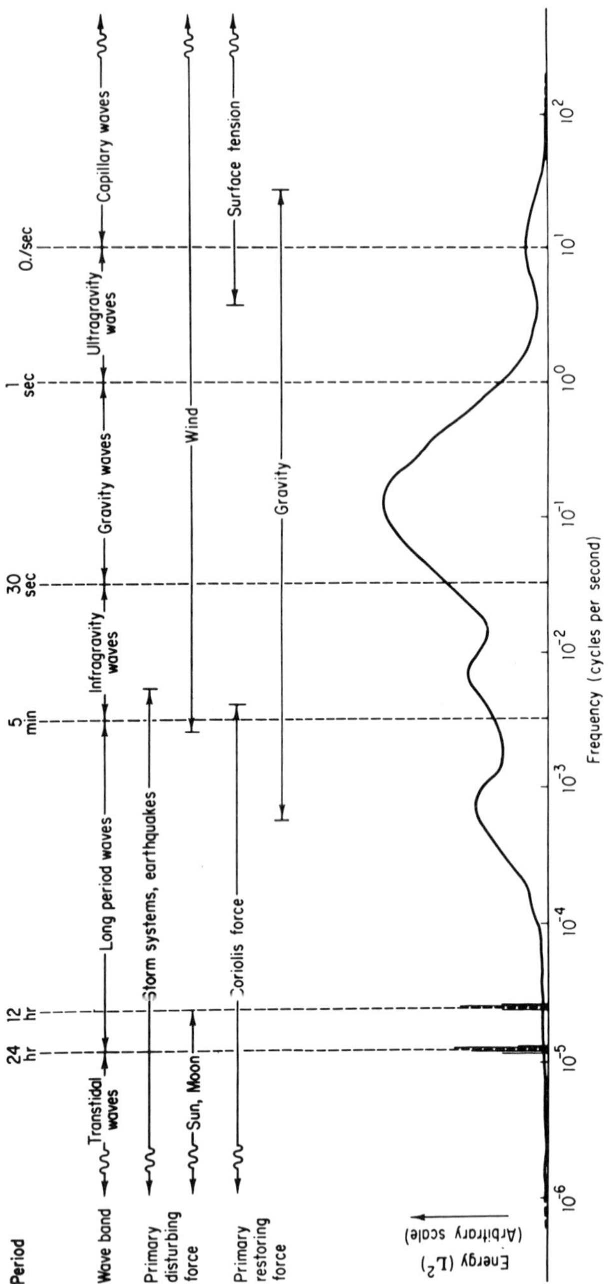

Fig. 5.4. Schematic (and fanciful) representation of the energy contained in the surface waves of the oceans—in fact, a guess at the power spectrum. (From B. Kinsman, *Wind Waves*, Englewood Cliffs, N.J.: Prentice-Hall, 1965.)

which are the waves in the wind-generating area and short-period waves with unsymmetrical slopes and steep crests. The characteristics of this confused sea depend on the wind speed and *fetch*—the length of the sea surface along the wind direction over which the wind blows more or less steadily. The other variety is the *swell*, which is a long and relatively symmetrical wave resulting from the wind waves after they leave the generating area and having a period between 10 and 30 sec. A mixture of sea waves with swell coming from a more distant storm is the result of the dispersive propagation of ordinary gravity waves. It is very difficult to describe the surface wave motion if one considers that over the ocean there are always one or more weather systems generating waves. The longest waves from each generating area travel with very little attenuation for considerable distances, and when they interact, a very complicated pattern results.

Beyond the ordinary gravity waves, periods between 0.5 to 5 min are present in the ocean. They are the *infragravity waves*. They seem to be generated by the interference taking place when two different wind wave trains arrive on the same shore at the same time. The beat frequency between those two wave trains produces a modulation in the sea level of an amplitude of a few centimeters. They are known as *surf beats*.

Longer-period waves are normally associated with meteorological disturbances. The pressure changes and wind shifts may generate standing oscillations over the width of the continental shelf of periods ranging from many minutes to a few hours. These waves have been correlated to Atlantic hurricanes and are explained as waves of the type of Lamb's *edge waves*, which are widely spaced wave crests oriented at right angles to the coastline. This type of wave can be generated by a shift in the wind associated with a passing meteorological front. The restoring force in wind waves and infragravity waves is gravity.

The long-period waves or transtidal portion of the spectrum involve waves in the period band of 5 min to 24 hr. Their wavelength is greater than the depth of the ocean. They behave like shallow water waves and are represented in the sea by storm surges, tsunami waves, and tides. The fact that the wavelength of these long waves is often comparable with the radius of the earth and the period comparable with a day makes it necessary to consider the effects related to the sphericity of the earth and its rotation. This applies particularly to tides. The waves are nondispersive and the speed of propagation or celerity depends only on the depth, h, of the ocean. Therefore, all long waves travel at the same celerity (200 m per sec in water 4000 m deep), which is also identical to the group velocity.

The storm surges are associated with atmospheric phenomena. The daily weather maps from oceanic regions and the climatic charts of the oceans clearly show that considerable variations of atmospheric pressure and wind intensity take place. Their periods range from less than a day to more than a season. The greatest perturbations are associated with storms and tropical disturbances that may have periods ranging from weeks to a fraction of a day. The pressure variations and/or changes in the wind stress acting over bodies of water excite the oscillation known as storm surges. The modes of oscillation have to be distinguished whether the periods are determined by the variation of the perturbating agents (pressure difference, wind stress) or by the properties of the body of water that is excited by the atmospheric agents. In the first class are included the surges of the running-wave type associated with intense storms, particularly the tropical storms (hurricanes, typhoons) which cause considerable damage to property and life. To the second class belong these particular oscillations known as *seiches*, which have been intensively studied in lakes.

The response of the sea level to the exciting forces depends on many factors. If atmospheric pressure changes slowly, the sea surface may reach a condition of quasi-equilibrium, in which case the sea level reacts as an inverted barometer, having an elevation where the atmospheric pressure is low and vice versa. In such a case, below a slow moving depression there is a mound of water which follows the pressure system and generates a rise in the sea level when the depression reaches the coast. However, the picture is not so simple, particularly in shallow water regions where water is piled up on the coast by the action of the winds. Tidal gauge records of most parts of the world show that the range of the sea level variation is greater (1) in areas of stormy weather, (2) along the edge of continents, and (3) in regions of broad and shallow continental shelves.

The wind presents higher frequency variations than the

atmospheric pressure. Its drag is more effective in shallow water, and this can be seen clearly in the vicinity of an estuary where its effect is more important than the integrated effect over a large adjacent oceanic region. The largest anomalies in the daily variation seldom exceed a meter.

The high-frequency surges which do most of the damage by inundating coastal regions are related to the transient state of the sea surface which is not in equilibrium with the disturbing forces. The surges generated by the tropical storms are of this type and are the most destructive.

Figure 5.5 shows the sea-level records at various ports along the eastern seaboard of the United States during the passage of a hurricane over the shallow, broad shelf. These records are typical of surges generated by hurricanes in those regions. It

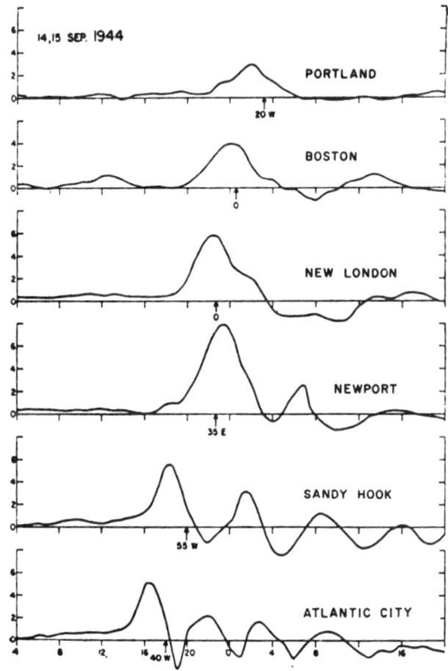

Fig. 5.5. Surge of September 14–15, 1944, along the Atlantic coast of the United States. Vertical scale is in feet, abscissa is in eastern standard time. The arrows indicate the time of nearest approach of the hurricane's center, distance in miles and direction of the tide gauge from the storm track (after Redfield and Miller, 1957, by courtesy of the American Meteorological Society). (From M. N. Hill, *The Sea*, Vol. I, New York: Wiley, 1962.)

has been possible to identify three successive stages: the *fore-runner* or initial surge which takes place as the hurricane is approaching and which may be an elevation or depression of the sea level depending on the relative motion of the hurricane with respect to the coast. Then follows a sharp rise in water level, which is the *hurricane surge* and happens when the center of the storm passes near the port. This surge does not last long (2.5 to 5 hr) and the amplitude can reach 3 to 4 m. The last surge is characterized by some *resurgences*, which are oscillations that occur after the passage of the storm.

The presence of the coast and the Coriolis effect may generate a type of free oscillation known as edge waves in which energy is trapped near the shore with the only possibility of flowing north or south. The resurgences shown in the tidal records of Fig. 5.5 have been suggested as being free edge waves traveling at the same speed of the storm center because the resurgence periods are almost similar to the resonant periods of the free edge waves moving at the speed of the disturbance.

It was mentioned before that the sea response to the exciting agents depends more on the properties of the body of water being perturbated. The geometric shape of an embayment or the width of a continental shelf may be such that their natural period of oscillation is tuned to the rapid fluctuating atmospheric agents or to some progressive wave coming from the deep ocean. In this case, that particular wave will be reflected back and forth from shore to shore or from shore to shelf edge. Once the exciting force has vanished, this resonant oscillation will go on for a time which depends on the dissipation suffered by the wave along its path.

This kind of resonant wave, generally known as *seiches*, can be recorded by the tide gauge stations as a damped undulation on the trace of the astronomical tide. Seiches can be generated in lakes, where they have been studied in considerable detail. The nature of the environment can put severe constraints to both seiches and progressive waves. They can be dissipated by gently sloping coasts and totally reflected by steep cliffs. Moreover, like other types of waves, they can be diffracted around obstacles and refracted by running into relatively shallow water or moving water. Seiching has created many unpleasant engineering problems, particularly in relation to harbor and port design. The standing oscillations inside the sheltered harbors or ports make it difficult for ships to remain comfortably tied to their docks.

The *tsunami*, also called, improperly, *tidal wave*, is generated by earthquakes and volcanic eruptions. It is a long, low wave propagating at the maximum speed for ocean depths ($c = \sqrt{gh}$) and losing energy density at a rate inversely proportional to the first power of the distance traveled. The coastal and bottom irregularities scatter considerably the energy of the tsunami out of the predicted paths from horizontal variations of $\sqrt{gh}$ into other directions and into other modes of ocean oscillation. Because of this scattering the tsunami created by the earthquakes or volcanic explosions, which lasted perhaps a few minutes, cause reverberations over the ocean for days with an estimated dissipation of 37 percent of its energy per day after the initial arrival.

During its propagation through deep water, the tsunami waves go unnoticed for an observer on board a ship because of its great wavelength and very small amplitude; however, when they enter into shallow water, the energy stored in the whole wavelength goes into increasing the height of the wave, which may reach values greater than 15 m (50 ft). The first indication of the arrival of a tsunami is an abnormal variation of the sea level; it may be either a crest or a broad trough depending on the kind of motion that originated the disturbance at the source point. A succession of smaller waves follows at intervals of 10–30 min and the associated rush of water may cause considerable loss of property and some loss of life. The Unimok tsunami, which occurred in the North Pacific on April 1, 1946, has been very well described. It consisted of a series of crests 60 cm (2 ft) high and 122 miles apart in the open ocean, propagating at about 750 km per hr (400 knots) so that the crest arrived at intervals of around 15 min. In Hawaii, the height of the water amounted to 17 m (55 ft) at Pololu Valley. The greatest loss of property and some lives took place in the city of Hilo, although the crest was half as high as it was in Pololu Valley. A ship lying offshore near Hilo did not notice any unusual waves. An international service for tsunami warning in the Pacific Ocean has been established in Honolulu under the auspices of the Inter-governmental Oceanographic Commission.

TIDES

The phenomenon of the tides is the periodic rise and fall of
the sea surface, which can be easily observed every day along
most coasts of the world. It is a forced oscillation, the forcing
function being the differential attraction of the moon and sun.
The normal tidal amplitude ranges between 30 cm to 2 m (1 to
6 ft), although considerably larger values may be found as a
consequence of particular oscillating characteristics of some
regions when subjected to the resultant of the gravitational
pull of the moon and the centrifugal reaction of the earth
around the common center of revolution (commonly known as
the *two-bodies problem*). To be more explicit, the problem can
be stated as follows: the earth and the moon constitute a system
in which the mutual Newtonian attractions regulate the monthly
motion around the system's center of gravity. The mass of the
earth is 81 times that of the moon, and the distance between
the centers of the moon and earth corresponds to 60 earth radii.
This relation places the common center of gravity inside the
earth. So, both the earth and the moon revolve in nearly circular orbits about this common center of gravity. Consequently,
observers on the earth feel both the attraction of the moon
and the centrifugal force of the earth. On the side of the earth
facing the moon, the gravitational attraction toward the moon
is somewhat greater than the opposing centrifugal reaction with
the result that the sea surface beneath the moon bulges out
toward the moon. On the other face of the earth, the centrifugal force away from the moon is stronger than the gravitational pull and the sea surface bulges out. There are, therefore,
two tidal bulges which will travel completely around the earth
once every lunar day.

The action of the differential force creating the tidal bulge
is exerted not only on the oceans, but also in the atmosphere
and in the solid earth. However, the response of these media
is different. For example, one of the great puzzles of tidal
theory has always been why the solar tide generated by the sun
is far greater in the atmosphere than the moon tide, whereas
the reverse is true in the ocean. The explanation lies in part
in that the natural oscillation period of the atmosphere is

about 12 hr, and therefore it resonates with the sun's gravitational pull.

Because of the moon-inclined orbit around the earth, the two daily high waters are generally not of equal height. This *diurnal inequality* goes through its own cycle, starting from zero when the moon crosses the earth's equator and following the declination cycle of the moon.

The other celestial body contributing to the generation of the tidal wave is the sun. Although its mass is considerably greater than the moon, it is much farther away and its tidal attraction on the earth is almost half as large as that of the moon. The simultaneous presence of the solar tides complicates considerably the pattern of lunar tides. On the other hand, it is possible to explain satisfactorily the presence of the large *spring tides* (occurring every half lunar month when the moon and the sun are in line with the earth and both tide-generating forces pull together) and the small *neap tides* (occurring a quarter of a lunar month later when the two celestial bodies are in quadrature).

The classic approach for the computation of such a complex wave as the tidal wave is similar to that used to describe the swell spectrum; that is, the wave may be considered formed by the addition of many periodic motions with the only simplification that a number of discrete components of definite frequencies determined from astronomic considerations are sufficient for such a computation, whereas a continuum of oscillations at all frequencies is necessary for the swell. The principal discrete components used for tidal prediction are M_2, S_2, N_2, K_2, K_1, O_1, P_1, M_1, and S_1 (the subscript indicates the approximate number of cycles per day). For the accuracy required in navigational safety the tides which are published in the *Tide Tables* for the different coastal regions of the world are computed with the use of between 37 to 62 harmonic components or partial tides.

In the previous analysis of the generation of the tidal bulge and its propagation across the ocean, nothing was said about possible constraints of the fluid and its container. It was assumed that the water of the ocean responded immediately to the forcing function—without inertia—and that the ocean surface was always perpendicular to the resultant of the force of gravity and the attractive forces of the moon and sun. This ideal condition is known as equilibrium tide and resembles the actual one mainly in the periodicity of the phenomenon. Under the assumptions of the equilibrium tide, the bulge should occur as the line joining the center of the earth and move along with the apparent diurnal motion of the sun and moon which would carry it around the earth in about 24 hr. This would imply that when the moon is at the celestial equator, the tidal wave must advance at a rate near 1000 mph at the equator and at decreasing values with increasing latitudes to zero at the poles. As the tidal wave is a shallow-water wave, the phase velocity is dependent on the water depth. The corresponding depth at the equator is 22.5 km to keep the equilibrium tide stationary under the sun as the earth rotates. As the average ocean depth is near 4 km, it is impossible for the tide to move freely at such a speed, but it is constrained by friction and the depth of the water. This explains why the equilibrium tide does not remain continuously beneath the sun and moon anywhere on earth but does cause the crest of the astronomic tide to lag behind sometimes for many hours the time that the moon and sun have crossed the local meridian.

That the ocean is always far from complete equilibrium with the generating forces can be seen from the properties of the long gravity waves which make up the tide. The sphericity and the rotation of the earth are responsible for the difference between the ideal wave and the actual tide. The main characteristics of the real astronomic tide are (1) horizontal motion is almost constant from the surface to the bottom and the water particles move back and forth at the wave frequency along straight horizontal lines in the direction of wave propagation (this velocity distribution is due to the fact that the wavelength is much greater than the water depth); (2) in the open sea, the earth rotation which manifests itself through the Coriolis force deflects all the straight motion of the particles to the right in the northern hemisphere (to the left in the southern hemisphere) and changes the horizontal displacements to ellipses in the horizontal plane. In restricted waters such as a canal or along the coast with the solid boundary at one side, the earth's rotation generates what is known as Kelvin edge wave. The process leading to the formation of this wave is as follows: the Coriolis acceleration tends to deflect the horizontal back-and-forth motion to the right (in northern hemisphere waters) but the solid boundary prevents the particle from moving in ellipses and keeps it moving in a straight line. For this to happen, the Coriolis force has to be balanced by a pressure gradient force normal to the coast which will increase the height of rise.

In some rivers or estuaries, because of the river discharge or hydraulic conditions, the rise of the water can be impeded to go upstream. In this case, the wave is held back in deep water until the continued rise of the tide outside the estuary deepens the water at the entrance sufficiently for the tide to go upstream. Once this condition has been reached, the wave, assuming a steep wall-like front over the total width of the river, propagates in such a fashion that following the passage of the wave, the water-level differences inside and outside of the estuary remains more nearly equal. The name *bore* (English) has been given to this conspicuous wave, and it is known to occur in many rivers, estuaries, and straits. The most impressive one is in the Tsien Tang Kiang estuary in China, which may be as much as 8 m (25 ft) high.

Another phenomenon of interest is the one related to the tides in coastal embayments. Progressive tidal waves entering embayments often produce standing waves owing to reflections at the head of the bay. The energy for the oscillation is derived from the ocean tides and is thought to be part of cooscillating systems in which the period is determined by the tide in the open sea, whereas the nature of the motion depends on the size and form of the enclosed water masses. In natural situations, the ideal standing wave is difficult to obtain because, normally, the heads of the bays do not properly reflect the impinging progressive wave. A mixture of progressive- and standing-wave types of tidal propagation results. This, together with irregular bottom topography, which may break the waves around shoals and accelerate in deep spots, complicates matters.

Besides the existence of yearly *Tide Tables*, which provide particular tidal data of many ports of the world for the use of the seamen and other operational purposes, the tidal information for scientific uses is given through the *tidal charts*. These charts give information about the geographical distribution of harmonic constants. Normally, the M_2 partial tides are represented, although the K_2, K_1, and O_1 tides are also sometimes given. The isolines plotted in the charts are corange lines (lines where the range of the tide has the same value) and cotidal lines (lines joining places where high water is occurring simultaneously). The most striking effect on a tidal chart is the existence of *amphidromic* points in which the range becomes zero and the cotidal lines converge. Figure 5.6 is a typical tidal chart for the Atlantic Ocean.

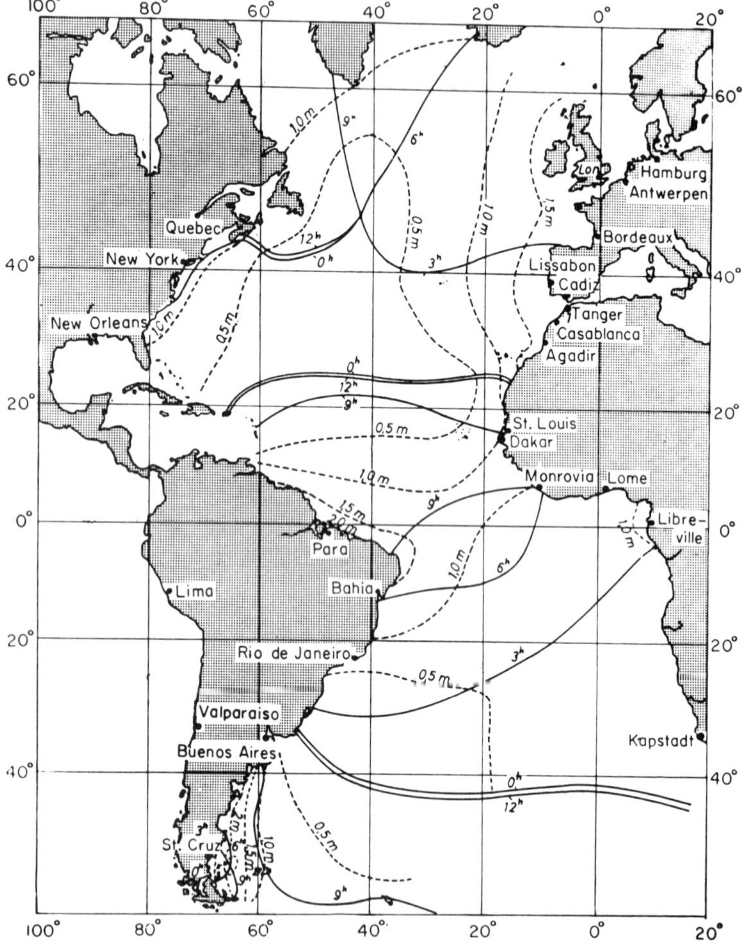

Fig. 5.6.   Cotidal and corange lines for constituent M_2. Numbers on the full lines give the time of high water after the moon's transit of Greenwich. The numbers on the broken lines give the range in meters (after Hansen, 1956). (From M. N. Hill, *The Sea*, Vol. 4, New York: Wiley, 1962.)

Because tides manifest themselves by the periodic rise and fall of the water, easily observed on most coasts of the world, one may tend to think about tides as a process related only to vertical displacements of the sea level. However, the horizontal motions are perhaps of more interest to coastal engineering. Tides and tidal currents are two expressions of the same phenomenon and are closely related. The direct measurement of tidal currents is more difficult and expensive than that of the tide and this explains the few observations available. The few long time-series observations made of tidal current at particular places suggest that tidal currents do not repeat themselves at one place as tides do, although the same inequalities described for tides are also present in tidal currents; like tides, tidal currents can be subjected to harmonic analysis. The predictions indicate that the current pattern consists of an ellipse only in areas with pure semidiurnal or diurnal tides. In areas with mixed tides, they become very complicated. Figure 5.7 shows the observed tidal currents in the whole column of water at the entrance of the English Channel.

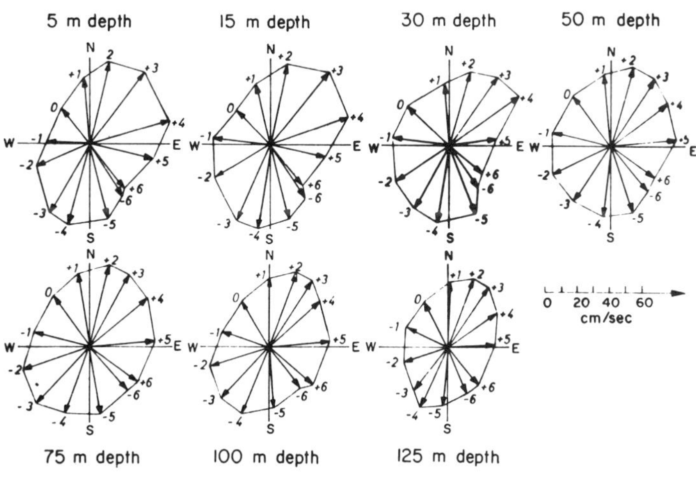

Fig. 5.7. Observed tidal currents at the entrance to the English Channel. The figures on the arrows give hours before or after the moon's transit of Greenwich. (From M. N. Hill, *The Sea*, Vol. 1, New York: Wiley, 1962.)

The strong tidal currents found often in particular coastal regions (up to 15 knots) are of special interest to navigation for several reasons. Of particular interest is the increase in speed of the ship relative to ground which may save valuable time. On the other hand, there is an undesirable phenomenon which may seriously affect the safety of vessels of even moderate size. This phenomenon is known as *refraction of wind waves on water currents.* If either sea or swell meets the strong current at a considerable angle, a very confused sea develops which lasts as long as the tidal current flows in that direction. Orbital motions become superimposed on the tidal current; hence, the wavelengths of the wind waves are shortened, the waves are steeper, and the wave crests are sharper. The sea becomes very irregular and ships are difficult to handle. The change of the tide produces dramatic effects because as soon as the tidal current ceases or changes direction, the sea surface becomes flat in a very short time. This phenomenon takes place in many regions of the world.

Tidal currents are of considerable importance for coastal engineering in tidal-flats environments because of the complicated sedimentation problem that generates the moving of considerable amounts of sediment during each tidal cycle. Moreover, very complicated sedimentation processes take place when river waters discharging in the coastal environment meet the combination of tidal currents and wind-wave action. They can render useless potential places for harbor construction.

Impressive amounts of energy are associated with tidal currents. This is now becoming one of the important ocean resources. Considerable efforts have been made to harness such an energy, and the first large tidal-current power station was built at the entrance of the Rance River near St. Malo in France, where the greatest tidal range of Europe is found. Around 500,000 kw will be generated by this hydroelectric power station.

Considerable advances have been made in the last few years in the understanding of global tides. The development of modern computers has made possible the undertaking of theoretical calculations of global tides for the true depths and boundaries of the world's oceans. The availability of tide gauges for use at moderate and deep depths in the ocean will permit the design of the proper network of observing stations that will check the theoretical tides.

GRAVITY WIND WAVES

A particular record of natural waves—heights against time— is a complicated one; and if study in detail is made, it is random in character and unpredictable. Another wave record taken at a later time will look completely different. However, it is obvious that the wave pattern in a stormy area will be different from that on a calm day; and in spite of the random character of the shape of the sea surface, a way has to be devised to describe such a difference. The logical approach is to look for the statistical properties of the wave system which are significant, and the most obvious one was to select the average wave height, period, and direction of travel. Although this description seemed to be satisfactory for some engineering problems, it does not provide all the information about the wave system that is needed for other purposes. For this reason, it was found that the most useful way of describing the wave pattern was by its *power spectrum,* which can be thought of as representing the distribution of energy over a scale of frequency. The different statistical models of sea waves attempted indicate that the best way to deal with this problem is to consider the sea surface as a random moving surface defined by a stationary Gaussian process. The study of sea waves has thereby developed into a combination of time-series analysis with a sort of statistical geometry consistent with the basic laws of hydrodynamics. Because of the little knowledge of the directional properties of sea waves, the practical methods of wave prediction, up to the present time, provide the *one-dimensional* spectrum which, in other words, is the frequency spectrum of a recorder which records the height of the sea surface above a reference on the sea bed. From this spectrum, the *significant wave height* and *significant wave period* required by engineers can be derived. For studies of ship motion in a seaway, the two-dimensional spectrum is needed.

Because no adequate body of theoretical knowledge is available to explain the generation of waves by wind, the present approach to wave prediction is of an empirical nature, although some theory is included. It aims to predict the wave spectrum from which the other parameters may be derived. All the methods presently available for wave forecasting begin by considering waves on deep water and assuming that the generation and development of a wave system depends on the wind *force* or speed, on how long the wind blows—*duration*—and on the *fetch,* the distance over which it blows. In the North Atlantic, the effective fetch, even during the strong winter gales, does not go over 600 nautical miles. A 40-knot wind blowing along this fetch could generate a 10-m (32 ft) wave.

It is more difficult to forecast wave conditions near the coast because of the influence of other factors, such as refraction due to changing depth of water and to tidal streams and dissipation of wave energy. For a particular area where many wave records are available, it is possible to improve the quality of prediction. From the various practical methods of wave forecasting, the most widely used at the present time is the one using Neumann's formula, known as the *Pierson, Neumann, and James Forecasting Method.*

The prediction of a wave spectrum, as indicated above, provides the information of the wave system in the generating area. The waves in this region are known as *sea.* When the sea waves leave the region of formation, they spread and propagate to other regions. There is a natural filtering process by which the shorter choppy waves disappear and the longer, undulating, and small-amplitude waves travel to great distances with the group velocity and little energy dissipation. These are known as *swells.* The first waves that arrive from a distant storm have a period of 20 sec. This first arrival is due to the higher group velocity. Wave groups of higher frequency travel more slowly and arrive later. It is possible to compute the distance of the storm that generated these waves by watching the arrival times of the different periods at one particular location. Heavy swell is responsible for severe damage to coastal installations and beaches. Swell conditions can be fairly well predicted, with enough anticipation, in oceanic areas where good meteorological information is available.

WAVES IN SHALLOW WATER

When deep water waves coming from the open ocean enter water of depth less than about half their wavelength, they feel bottom and begin to undergo a transformation which terminates with the obliteration of the wave on the beach and the release of their original wave energy in different processes in the nearshore region. The first transformation is the refraction by decreasing depth. Waves approaching the coast at some angle tend to become oriented to the bottom contours because of wave refraction. The part of the wave front that reaches the shallower depth first advances more slowly than the part traveling in deeper water, and the wave front bends in such a way as to fit the shoreline and break up on it almost at once. This refraction process continues as long as there is some bottom relief and a definite wave to be refracted. The waves will be focused into shallower regions so that the amount of energy at work in those areas, such as submerged ridges, exposed capes, and shallow beaches, will increase considerably. The opposite effect is experienced in deeper regions where the waves diverge and the wave activity will be minimal. A clear picture of the refraction processes and intensity of wave activity can be obtained by looking at the trend of the *ray orthogonals,* which are lines normal to the wave crests. Where they converge, the wave activity will be greatest. The refraction effects of focusing and diverging depends on the incidence angle of the incoming deep water waves; a small change in this angle may result in dramatic changes in wave intensity at certain points on the coast. This stresses the importance of knowing the wave climate and its variability before undertaking coastal modifications. Figure 5.8 illustrates the refraction process. Refraction can also take place when incoming waves meet tidal currents, which may be strong along the coast. This type of refraction was mentioned earlier in this chapter. This property of waves refracting and dissipating energy in hydraulic discontinuities has been applied in the design of hydraulic and pneumatic breakwaters for harbor protection.

The second transformation is the change of the deep water wave to the shallow water wave as the relative depth decreases. This implies that the celerity decreases and depends on the depth, h (due to the relation $c = \sqrt{gh}$); the wavelength also decreases and the period is kept constant. The energy is then concentrated in a shorter wavelength and the waves become shorter, higher, and steeper. The symmetry, with respect to the undisturbed sea surface, is lost; the troughs become broader and the

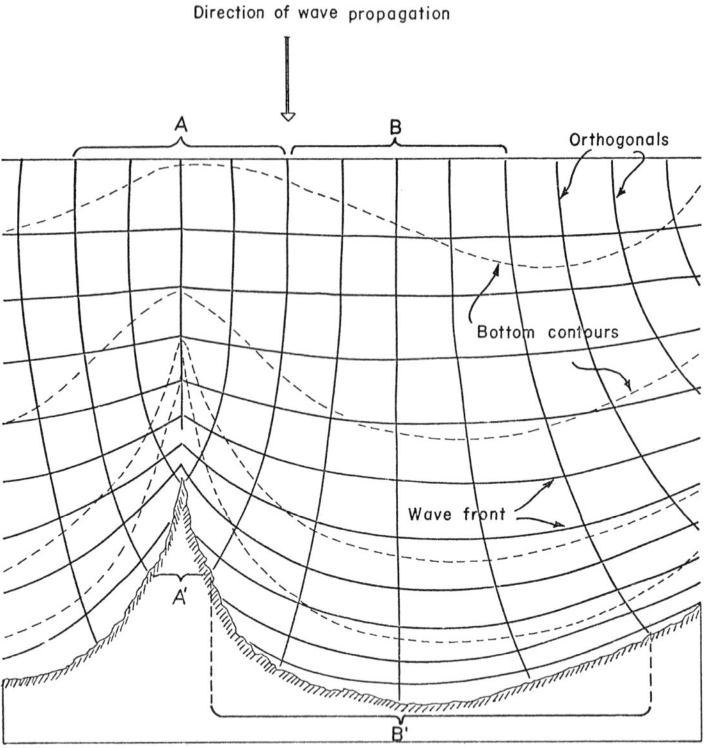

Fig. 5.8 Wave refraction diagram showing effects of bottom topography on wave propagation. Energy concentrates at submarine ridge (A′) and spread at the deeper indentation B′.

crests steepen, thereby resembling solitary peaks. If the depth decreases slowly and evenly, the wave will break as a *spilling wave* in such a fashion that turbulent water is spilled out from the crest and the wave will lose energy as it goes through the surf zone until completely dissipated. If the depth diminishes rapidly, the spilling stage never develops and the energy is then dissipated in a single forward plunge. This type of wave is the *plunging breaker*.

The surf zone is the region where drastic changes in the wave shape take place; nonlinear processes become very important and create formidable problems for theoretical treatment. Depending on the beach slope, breaking may not take place, but reflection of the wave back to open sea may occur. It is difficult to define the slope values which may reflect waves or create conditions for spilling or plunging types of breakers. As a rough guide, it may be said that a plunging wave develops where the undisturbed depth is a little greater than the height of the breaker. Spilling breakers can be initiated in much deeper water.

Water waves can be diffracted into the shadow zone behind steep-sided obstacles provided the relation between the size of the obstacle and the wavelengths are fulfilled. Because of the nature of the obstacles present at sea, it is difficult to differentiate diffractive effects from refractive ones.

In the Pacific Ocean, coral reefs of the fringing type around atolls rise as nearly vertical walls. In some places, the dimensions of the submerged portion of the reef are about one-half wavelength high, and diffraction phenomena may predominate. The Caribbean Islands in the North Atlantic are natural structures which may develop diffraction-grating effects into the Caribbean Sea where the long-crested swell of the trade-wind zone passes through the gaps between the islands. However, it must be stressed that diffraction phenomena are well developed where the structures have nearly vertical slopes for a depth as great as one-half wavelength.

Diffraction effects may become a real nuisance in coastal installations, particularly in the design of seaports, breakwaters, jetties, and other constructions that present an obstacle to the propagation of the wind waves. Considerable research has been done on this problem, and different models have been tested with theoretical results. There exist mathematical solutions of some particular problems which are adaptations of water waves for the case of water of constant depth and for impermeable rigid structures to similar problems of acoustic and light-waves physics. Diffraction produced by semi-infinite breakwater, single breakwater gap, detached breakwater, or other obstacles has been studied and there are graphic means of predicting diffracted wave heights in the lee of breakwaters and similar structures or natural obstacles. A difficult problem to solve because the linear theory cannot be applied is the *Mach-stem* type of reflection where energy is concentrated in certain conditions and waves can reach the geometric shadow by swinging around a curved breakwater. Figure 5.9 shows how incoming waves are diffracted by a breakwater.

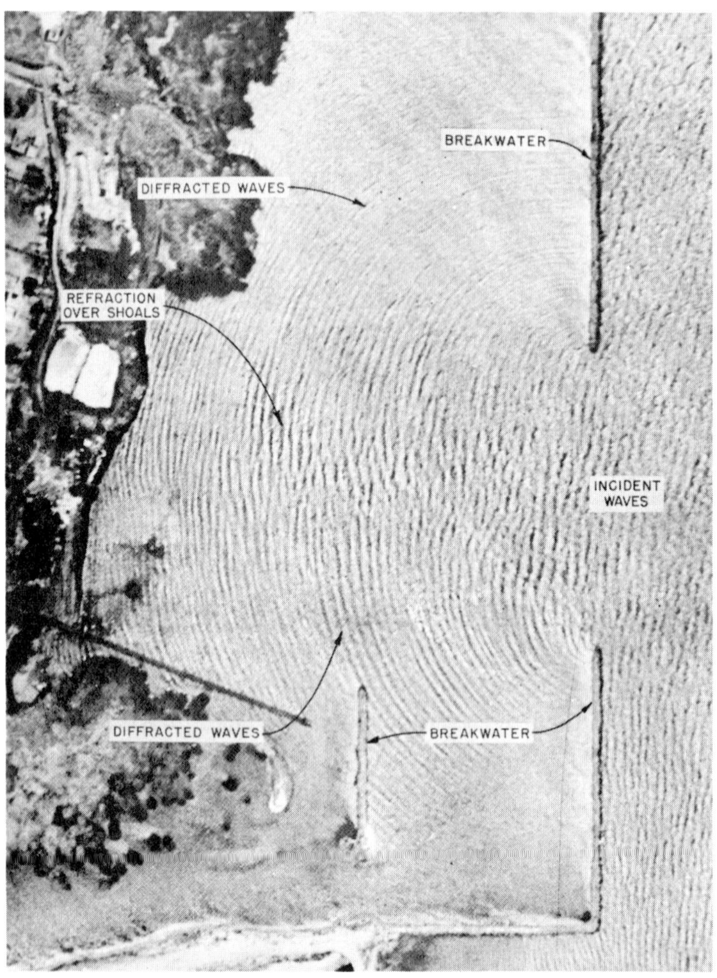

Fig. 5.9.   Diffraction of waves at a breakwater. (From J. W. Johnson, *Proceedings of the 2nd Conference on Coastal Engineering*, American Society of Civil Engineers, 1951.)

The final transformation of the wave is its dissipation in the surf zone. In this region, the waves drive increasing quantities of water forward toward the shore. This water becomes the main source of a very important nearshore process, the *nearshore circulation system*. This problem will be dealt with in Chapter 7.

WAVE FORCES

When a structure is placed in the ocean, forces induced by the motions of the water are applied to it. Of particular interest because of their magnitude are the forces generated by waves. The failures experienced by many massive structures provide dramatic examples of the magnitude of such forces. The theoretical determination of the forces exerted by waves on structures is very difficult. In most cases, it was necessary to devise semi-empirical methods for the adaptation of more classic hydraulic problems to the special conditions where wave motion is present.

The general approach to this problem is to formulate the equation of the forces acting on rigid submerged bodies in unsteady flow. The two approaches to solve this problem are (1) formulation of the boundary-value problem for a potential flow with the addition of a semiempirical equation for the drag forces, and (2) the semiempirical type of equations which have been employed successfully in undirectional flow. Both methods require the knowledge of water particle motion and coefficients to be empirically determined. The theoretical value of the equations is limited to the degree of accuracy with which the orbital motion of the waves can be predicted; and, in this area, the problem is still far from being satisfactorily solved. The Stokes' wave theory is applicable to either low or finite height in deep water, and for waves of infinitesimal height (in relation to wavelength and water depth) in relatively shallow water. Very little data are available to check the accuracy of the theory. The cnoidal wave theory is valid only in relatively shallow water, and no satisfactory theory exists which provides the relation between water particle velocities and accelerations related with the wave motion in the breaking zone. At present, there are at least twelve available wave theories which can be chosen for wave motion description. The selection of the theory which may fit best with the particular conditions has to be made on the basis of the relative depth conditions for which a particular theory was developed. However, most of the theories are difficult to apply and the engineering design parameters are also difficult to calculate. Relative evaluation of the available different theories has been made for the case of boundary-value problems in potential flow. This evaluation confirms the limitations stated.

INTERNAL WAVES

In addition to surface waves, there is a kind of wavelike oscillation within the ocean water known as *internal waves.* This mode of motion is possible because of the layered stratification of the water mass. Internal waves are present in all oceans, probably most bays and lakes, and vary widely in amplitude, period, and depth. They can be detected by the oscillations of the isotherms since temperature measurements at sea are easily accomplished. Their amplitudes may exceed those of surface waves, but their speed is smaller and have periods of many minutes. The existence of this type of wave has no effect upon the vertical displacements of the sea level, but it may mask the orbital currents of the tide because of its large and variable horizontal velocities throughout the body of the fluid.

Gravity is the restoring force, but the difference with the sea surface is that the buoyant forces arising from the increase of density with depth are also added. Therefore, the layered density structure within the water mass that tends to oscillate meets a configuration more stable than a free surface and it rebounds from any displacement in a manner similar to a spring released from tension. The frequency of oscillation depends on the stability of the water and is the Väisälä–Brunt frequency, explained in Chapter 3. In a two-layer ocean, the maximum amplitude is present at the boundary of two layers and decreases linearly with distance upward and downward. In a continuous density gradient system, as is the case in the sea, the internal waves are much more complicated, and there are indications that an infinite number of internal wave modes may exist in the ocean. If energy is to be transported as a wave, the frequency of each internal wave must necessarily be greater than the Väisälä–Brunt frequency at a given depth.

It has been determined that the celerity of internal waves is smaller than that of surface waves; therefore, for waves of the same period, they have much shorter wavelengths. A first-mode internal wave, which resembles the wave propagating in a two-layer ocean, of 24-hr period moves at about 5 m per sec and has a wavelength of about 430 km. This is very large for an oceanic internal wave; wavelengths of less than 100 km are more typical of higher-mode 24-hr internal waves. The shorter the periods, the shorter the wavelengths; the most rapid ones, with periods of about 10 min, have wavelengths of around 100 m.

The origin of the internal waves in the ocean has not been well determined. The high-frequency oscillations may, in part, be of atmospheric origin. It has been suggested that internal waves dissipate considerable amounts of tidal energy when the tide moves over bottom roughness of small scale (10 to 50 km) and that the continental shelf radiates internal waves of tidal period.

Besides detecting the waves by measuring the temperature at various depths and determining the rise and fall of the isotherms, it is possible to have some surface manifestations which can be seen with the naked eye. For instance, when internal waves are present below some surface indicators such as a thin layer containing plankton or silt, long parallel bands moving slowly are seen at the surface of the sea. These bands have been explained as a result of converging motion at the sea surface over the trailing slope of the internal wave. The slicks appear in calm weather because the concentrations of surface contamination prevent the formation of ripples under light winds. It is also possible to see the effects of internal waves by watching a submarine hovering in near-perfect trim beneath the surface. As the stratified layers move up and down along a strong density interface, so does the submarine. If it is cruising, the changes in buoyancy going through crests and troughs make it difficult to keep a constant depth solely by adjusting the angle of the diving planes. The classic example of the effects of internal waves is the phenomenon known as *dead water.* This may happen in regions where fresh surface layers produced by river discharge or melting of ice in polar regions extend over salt water creating a very strong interface. These fresh layers may be so thin that a moving ship may generate an internal wave that, being so shallow, will affect the progress of the ship.

The propagation of sound in the sea is affected by the presence of internal waves; it creates a problem to activities that make use of sound measurements. Of particular importance are the implications on antisubmarine warfare because of the intensive use of sonic devices. The refraction and sound-focusing effects which are produced by the undulating thermocline consist of horizontal changes in the sound intensity of considerable magnitude and intermittent zones of high and low intensity.

SUMMARY

Waves in the ocean are a periodic phenomenon characterized by a fast propagation of a disturbance without transport of water. The most common waves are the surface waves generated by atmospheric action, particularly the wind. The relative depth h/L (depth divided by wavelength) is the main parameter to determine wave behavior. For deep water waves $(h/L > \frac{1}{2})$ energy moves at group velocity which is half of the celerity, or phase speed. For shallow water waves $(h/L < \frac{1}{20})$, energy moves at group velocity equal to phase velocity. Because of the value of the mean ocean depth, all long-period waves, such as tsunamis and tides, propagate like shallow water waves.

Enclosed bodies of water, natural indentations along the coast, or man-made facilities display natural water oscillations known as seiches. Their periods depend on the geometric configuration, and extreme values can be reached when the period of the perturbating agent approaches the natural period (resonance conditions).

Wave phenomena can be predicted with a different degree of accuracy depending on the band of the spectrum to be considered. Tides are well known; sea and swell forecast are continuously improving; storm surge is the object of considerable research. Internal wave prediction is still far from desirable.

Wave phenomena are of particular interest to the oceanographic engineer in the following areas:

1. Wind waves which create a formidable problem for the motions and forces generated in the different kinds of platforms—ships, buoys, and towers—and the noise in the sensors measuring oceanographic phenomena.

2. Wave dissipation along the shores, where most recreation activities take place and where most of the man-made facilities are built, is a very complex phenomenon. Wave forces on the structures and sediment transport, because of the nearshore circulation system, are typical examples of the processes taking place along this interface.

6

The Chemistry and Biology of the Marine Environment

The simple statement that sea water is an electrolyte or salty solution in which some major constituents are present in relatively constant proportion does not provide a fair indication of the scope of the problem faced by the marine chemist. His challenge is to understand the chemistry of a natural, large-scale system; or, expressed in other words, the oceans as a chemical system. Therefore, the chemical oceanographer has to deal with a solution of gases as well as several inorganic salts, including divalent sulfates and carbonates and a high concentration of sodium chloride, over a range in hydrostatic pressure of around 1000 atm and a temperature range from freezing up to $30°C$. It is a system which is continuously exchanging matter and energy through the natural boundaries and where concentrations of the solutes are always being modified by the activities of the organic environment (living organisms). The solution is constantly stirred by the natural turbulent state of motion of the water, and the molecular diffusion processes are overshadowed by eddy processes.

In Chapter 3, when discussing the physical properties of sea water, emphasis was placed on the anomalous character of the water substance with unusually high heat capacity, latent heat of vaporization, boiling point, density, and dielectric constant, among other unusual properties. On the basis of this anomalous behavior, it was suggested that the well-known water molecule formula H_2O does not adequately represent the water molecule in the marine environment. It is likely that water molecules are present in clusters of two or three joined by a variety of chemical bonding processes (polymerization) so that their effective average molecular weight is much greater than 18.

The chemistry of aqueous solutions for single, univalent salts at low concentration is fairly well understood. The research related to the above solutions required to make conductivity measurements accurate to 0.01 percent in turn implies great purity of the salt water uncontaminated by carbon dioxide and precise temperature control to 0.001° C. This high degree of accuracy in the chemical measurements and purity of reagents is a common requirement in most chemical determinations.

In oceanic areas far from the mouths of rivers, the relative concentration of the major constituents is constant. These stable major solutes amount to 11 dissociated salts that make up over 99.9 percent of the total dissolved salts in sea water. These are the cations sodium, magnesium, calcium, potassium, strontium, and undissociated boric acid, and the anions chlorine, sulfate, carbonate, bromine, and fluorine. The total dissolved salts in sea water, as represented by a single solute defined by salinity, was explained in Chapter 3.

To understand the chemistry of the ocean, which is a multi-constituent system, it is necessary to have more accurate data and more realistic theory concerning the effects of salts on the dissociation constants of weak acids and ion pairs (such as carbonic acid, boric acid, magnesium sulfate, and dissolved calcium carbonate) and on the solubility of gases and solids, particularly calcium carbonate. It is also necessary to determine the effects of pressure and temperature on chemical equilibria in single salt and eventually polyelectrolyte solutions. The effects of pressure on solubilities, conductivity, equilibrium constants, activity coefficients, and other properties must be well investigated to understand the thermodynamics of sea water.

An example which clearly illustrates the complexity of the ocean as a chemical system and the need for a broad approach to marine chemistry is the attenuation of sound in the ocean. It has been demonstrated that the sound absorption coefficient in sea water is considerably greater than that observed in fresh water. Experiences carried out in artificial sea water showed that this unusual effect was also present in several of the divalent sulfates, and it was necessary to devise a four-state model of dissociation to account for the observations at atmospheric pressure. The four-state concept has been supported by recent measurements on both conductivity and sound absorptions at different pressures.

In summary, the research activity in the chemistry of sea water must be undertaken along several fronts which include studies of the following: single and polyelectrolyte solutions, the theory of high concentration solutions, and the equation of state of sea water. These studies involve measurements of the effect of the environmental elements of temperature and pressure on conductivity, activity coefficients, osmotic pressure, partial molal volumes, solubilities, reaction rates, dielectric constants, density, viscosity, nuclear magnetic resonance, electrochemical potential, and sound absorption.

COMPOSITION OF SEA WATER

It is somewhat difficult to specify the composition of sea water in its natural environment if one takes into consideration the different processes that are constantly changing the concentration of the different solutes. These processes are responsible for the variability of the distribution and concentration of the elements in the ocean waters. They consist of river runoff discharging its waterladen sediment in the sea and chemical reactions at phase discontinuities (atmosphere–hydrosphere, sediment–hydrosphere, and the biological activity which is responsible for the most dramatic compositional changes and regulation of the abundances of certain important elements).

The considerations which outline the characteristics of the oceans as a chemical system were well postulated by Forchhammer in 1865 in the following way: "Thus the quantity of the different elements in sea water is not proportional to the quantity of elements which river water pours into the sea, but inversely proportional to the facility with which the elements in sea water are made insoluble by general chemical or organochemical actions in the sea."

In view of the characteristics of the marine environment, it seems adequate to describe the composition of the sea water as it occurs in its natural environment in the following way: (1)

average composition of ocean water, (2) relative reactivities of
the elements based on the concept of residence time of the
elements in the ocean, and (3) the spatial and temporal distributions of the elements.

Table 6.1 gives the average composition of sea water in the
ocean. This table has some limitations because many of the

TABLE 6.1 Geochemical Parameters of Sea Water*

Element	Abundance, mg/liter	Principal species	Residence time, yr
H	108,000	H_2O	
He	0.000005	He (g)	
Li	0.17	Li^+	2.0×10^7
Be	0.0000006		1.5×10^2
B	4.6	$B(OH)_3$; $B(OH)_2O^-$	
C	28	HCO_3^-; H_2CO_3; CO_3^{2-}; organic compounds	
N	0.5	NO_3^-; NO_2^-; NH_4^+; N_2 (g); organic compounds	
O	857,000	H_2O; O_2 (g); SO_4^{2-} and other anions	
F	1.3	F^-	
Ne	0.0001	Ne (g)	
Na	10,500	Na^+	2.6×10^8
Mg	1350	Mg^{2+}; $MgSO_4$	4.5×10^7
Al	0.01		1.0×10^2
Si	3	$Si(OH)_4$; $Si(OH)_3O^-$	8.0×10^3
P	0.07	HPO_4^{2-}; $H_2PO_4^-$; PO_4^{3-}; H_3PO_4	
S	885	SO_4^{2-}	
Cl	19,000	Cl^-	
A	0.6	A (g)	
K	380	K^+	1.1×10^7
Ca	400	Ca^{2+}; $CaSO_4$	8.0×10^6
Sc	0.00004		5.6×10^3
Ti	0.001		1.6×10^2
V	0.002	$VO_2(OH)_3^{2-}$	1.0×10^4
Cr	0.00005		3.5×10^2
Mn	0.002	Mn^{2+}; $MnSO_4$	1.4×10^3
Fe	0.01	$Fe(OH)_3(s)$	1.4×10^2
Co	0.0005	Co^{2+}; $CoSO_4$	1.8×10^4
Ni	0.002	Ni^{2+}; $NiSO_4$	1.8×10^4
Cu	0.003	Cu^{2+}; $CuSO_4$	5.0×10^4
Zn	0.01	Zn^{2+}; $ZnSO_4$	1.8×10^5
Ga	0.00003		1.4×10^3
Ge	0.00007	$Ge(OH)_4$; $Ge(OH)_3O^-$	7.0×10^3

TABLE 6.1 Continued

Element	Abundance, mg/liter	Principal species	Residence time, yr.
As	0.003	$HAsO_4^{2-}$; $H_2AsO_4^-$; H_3AsO_4; H_3AsO_3	
Se	0.004	SeO_4^{2-}	
Br	65	Br^-	
Kr	0.0003	$Kr\,(g)$	
Rb	0.12	Rb^+	2.7×10^5
Sr	8	Sr^{2+}; $SrSO_4$	1.9×10^7
Y	0.0003		7.5×10^3
Zr			
Nb	0.00001		3.0×10^2
Mo	0.01	MoO_4^{2-}	5.0×10^5
Tc			
Ru			
Rh			
Pd			
Ag	0.0003	$AgCl_2^-$; $AgCl_3^{2-}$	2.1×10^6
Cd	0.00011	Cd^{2+}; $CdSO_4$	5.0×10^5
In	<0.02		
Sn	0.003		5.0×10^5
Sb	0.0005		3.5×10^5
Te			
I	0.06	IO_3^-; I^-	
Xe	0.0001	$Xe\,(g)$	
Cs	0.0005	Cs^+	4.0×10^4
Ba	0.03	Ba^{2+}; $BaSO_4$	8.4×10^4
La	0.0003		1.1×10^4
Ce	0.0004		6.1×10^3
Pr			
Nd			
Pm			
Sm			
Eu			
Gd			
Tb			
Dy			
Ho			
Er			
Tm			
Yb			
Lu			
Hf			
Ta			
W	0.0001	WO_4^{2-}	1.0×10^3

TABLE 6.1 Continued

Element	Abundance, mg/liter	Principal species	Residence time, yr.
Re			
Os			
Ir			
Pt			
Au	0.000004	$AuCl_4^-$	5.6×10^5
Hg	0.00003	$HgCl_3^-$; $HgCl_4^{2-}$	4.2×10^4
Tl	<0.00001	Tl^+	
Pb	0.00003	Pb^{2+}; $PbSO_4$	2.0×10^3
Bi	0.00002		4.5×10^5
Po			
At			
Rn	0.6×10^{-15}	$Rn\,(g)$	
Fr			
Ra	1.0×10^{-10}	Ra^{2+}; $RaSO_4$	
Ac			
Th	0.00005		3.5×10^2
Pa	2.0×10^{-9}		
U	0.003	$UO_2(CO_3)_3^{4-}$	5.0×10^5

*From M. N. Hill, *The Sea*, Vol. II, New York: Wiley-Interscience, 1963.

values were obtained from a single set of analyses of surface waters, which very likely are not representative of the whole marine environment. The greatest variations in the concentration of elements takes place in the upper 100 m or so where the intense biological activity produces the large-scale fractionations of certain materials. Certain elements such as iron and silicon, which play an important role in biochemical cycles, can have differences in concentrations of the order of 10^3 or higher in two different water masses.

The composition of the sea water given by the concentration of each element is important for certain initial considerations. However, if one wishes to gain an insight into the behavior of the chemical system, such as the study of equilibrium reactions in the marine environment and the stability of various dissolved species, it is convenient to know the forms in which the reacting elements are present in sea water. It also facilitates the postulation of coherence in geochemical behavior. This information is given in the third column of Table 6.1.

The last column provides the information of the *residence times*. This concept gives an indication of the capacity of the element for sustaining chemical reaction in the ocean. It measures the passage time of an element in the ocean waters by assuming an ocean in steady state in which an amount of certain element introduced per unit of time is balanced by a similar amount deposited in the sediment. Therefore, the residence time can be thought of as the average time the element remains in the ocean water before being removed by some precipitation process. The elements having long residence times, the alkali metals and alkaline earths, are characterized by little reactivity in sea water. On the other hand, some elements shown in the table have residence times under 1000 yr; this interval is almost similar to the mixing times for oceanic-water masses.

The presence of isotopes of some elements not shown in the table affects some of the physical properties of sea water and may be responsible for some of the observed anomalous behavior. The fractionation of those isotopes by geochemical or biological processes has been used to explain the nature of the mechanisms involved. The isotopic compositions of the oceanic waters are strongly affected by the previous geophysical processes, such as condensation, evaporation, freezing, and melting. Both hydrogen and oxygen have a pair of stable isotopes which are significantly fractionated in a similar way in oceanic waters. The isotopes of nitrogen and sulfur are useful in extending the knowledge of chemical reactions in the marine environment.

The most common way to report the average composition of sea water is shown in Table 6.2. It lists the major solutes (often called the conservative elements) which make up over 99.9 percent of salinity. The concentrations given in this table are for water of chlorinity 19 ppm. The minor elements exist in micromolar and submicromolar concentrations; however, they can participate and affect some of the more important inorganic and biochemical reactions of the ocean waters. They are listed in Table 6.1.

DISSOLVED GASES IN SEA WATER

The oceanic waters are, in general, rather well aerated. Atmospheric gases acquired through the air–sea interface are dissolved within the water mass. The physical processes of the

TABLE 6.2 Composition of Sea Water—Concentration of Constituents in Sea Water Having a Chlorinity of 19%

	g/kg	g/unit of chlorinity)
Chloride, Cl^-	18.890	0.99894
Sodium, Na^+	10.560	0.5556
Magnesium, Mg^{++}	1.273	0.06695
Sulfate, SO_4^{--}	2.649	0.1394
Calcium, Ca^{++}	0.4104	0.02106
Potassium, K^+	0.380 mg/kg	0.02000
Carbon (as HCO_3^- or CO_3^{--})	28	0.00735
Bromide, Br^-	65.9	0.00340
Strontium, Sr^{++}	8.1	0.00070
Boron, as H_3BO_3	4.6	0.00137
Silicon, as silicate	0.01–4.5	
Fluoride, F^-	1.4	0.00007
Nitrogen, as NO_3^-	0.01–0.80	
Aluminum, Al^{+3}	0.5	
Rubidium, Rb^+	0.2	
Lithium, Li^+	0.1	
Phosphorus, as PO_4^{-3}	0.001–0.1	

Trace Elements present in concentrations of 1 to 50 μg/kg: barium, iodine, arsenic, iron, manganese, copper, zinc, lead, selenium, caesium, uranium.

Trace elements present in concentrations of less than 1 μ/kg: molybdenum, thorium, cerium, silver, vanadium, lanthanum, yttrium, nickel, scandium, cobalt, cadmium, mercury, gold, tin, chromium, radium. (From John F. Brahtz, *Ocean Engineering*, New York, Wiley, 1968.)

atmosphere and ocean interchanging gases take place continuously depending on the partial pressure of the particular gas in both media. Gas goes from the medium with higher partial tension to the other medium until equilibrium is reached. A basic assumption in chemical oceanography is that every parcel of ocean water has been, at one time, at the surface and in contact with the atmosphere when it became saturated with the atmospheric gases. Being a participant of the oceanic circulation, the parcel carries the gases with it; and by processes of diffusion and advection, the gases are distributed throughout the entire body of water. When away from the surface interacting layer, some of the gases, particularly oxygen, may change the concentration considerably because of biological processes taking place within the water mass. The amount of the gases that dissolve in the surface waters depends on the solubility coefficients which are functions of salinity and temperature.

By far, the most abundant gases are nitrogen, oxygen, and argon; and there are traces of the noble gases, carbon dioxide, and a variety of others introduced by man's activities. Because its concentration is easily determined and because of its involvement in biological processes, the dissolved oxygen distribution and its variations have been extensively studied by oceanographers.

The ratios of the different gases in the air are not the same as the ratios in air-saturated sea water because their solubilities in water are different. Also, the absolute concentration in both media is different, being much smaller in the water than in the air (with the exception of carbon dioxide). Thus, the amount of oxygen in the atmosphere is around 210 ml per liter while its concentration in sea water is between 5 and 10 ml per liter. From these values, it can be seen that the oxygen requirements of air-breathing animals are easily met, while the marine organisms must have efficient respiration organs and/or lower metabolic requirements for extracting oxygen from the low concentrations found in sea water. The contrary is true for carbon dioxide, as its concentration in the ocean waters is about 50 ml per liter while in the atmosphere it is 0.3 ml per liter. This difference reflects some of the differences in which marine plants and land plants carry on their photosynthetic activities.

The concentrations of oxygen, carbon dioxide, and nitrogen can also be altered by biological processes. Photosynthesis and respiration influence the concentration of the first two gases in the same way as in the terrestrial environment. In the lighted upper layers of the sea, the *euphotic layer*, light energy and carbon dioxide are present in such an amount as to let the photosynthetic plants carry on the generation of organic matter from the inorganic environment—primary production. Carbon dioxide is removed from the water and oxygen is generated. At the same time, respiratory processes work in the opposite way.

Because of the characteristic of the photosynthetic processes in the marine environment, the concentrations of dissolved oxygen vary widely from the concentrations that might be expected from the solution of air. This is clearly seen in Fig. 6.1. In the upper layers of the sea (between 25 to 50 m), the concentrations of oxygen are up to twice the values to be obtained if only saturation from the air takes place. This difference is evidently the result of photosynthesis. The oxygen minimum is found at middepths and increases with depth. This increase is attributed to water renewal by waters of high density that sunk in high latitudes. These surface waters were initially characterized by high oxygen and relatively low organic contents. This convective circulation was discussed in Chapter 4.

The changes in nitrogen concentration are due to different biological processes related to bacterial activity. They are nitrogen fixation and denitrification. The first is the formation of nitrogen compounds from the free dissolved nitrogen, and denitrification is the opposite process in which oxidized nitrogen compounds, nitrate and nitrite ions, are reduced to free nitrogen. This kind of process appears to be restricted to certain oceanic regions where water is not renewed and stagnation occurs. In these areas, the dead bodies of animals and plants fall to the bottom and are subjected to bacterial attack. At the beginning, bacterial respiration takes place at the expense of dissolved oxygen; after the oxygen has been consumed (no renewal takes place), denitrification begins. The classic oceanographic region where denitrification can proceed is the Black Sea, which has a depth of about 2000 m and no oxygen is available below depths between 150 and 250 m. Anoxic conditions are more common in the bottom sediments. Here, the oxygen can be supplied only by slow diffusion processes; the available oxygen is not enough to take care of the rich organic matter being deposited and the sediments become devoid of oxygen. This particular environment is of particular engineering interest because of the anaerobic corrosion of iron in soil.

Carbon dioxide in the ocean has been extensively studied because of its bearing on the regulation of the pH. The greater part of the dissolved carbon is present as carbon dioxide, carbonic acid, bicarbonate ion, and carbonate ion, which together with the hydrogen ion constitute the carbonate system in the ocean. The equilibrium of this system works in such a way as to

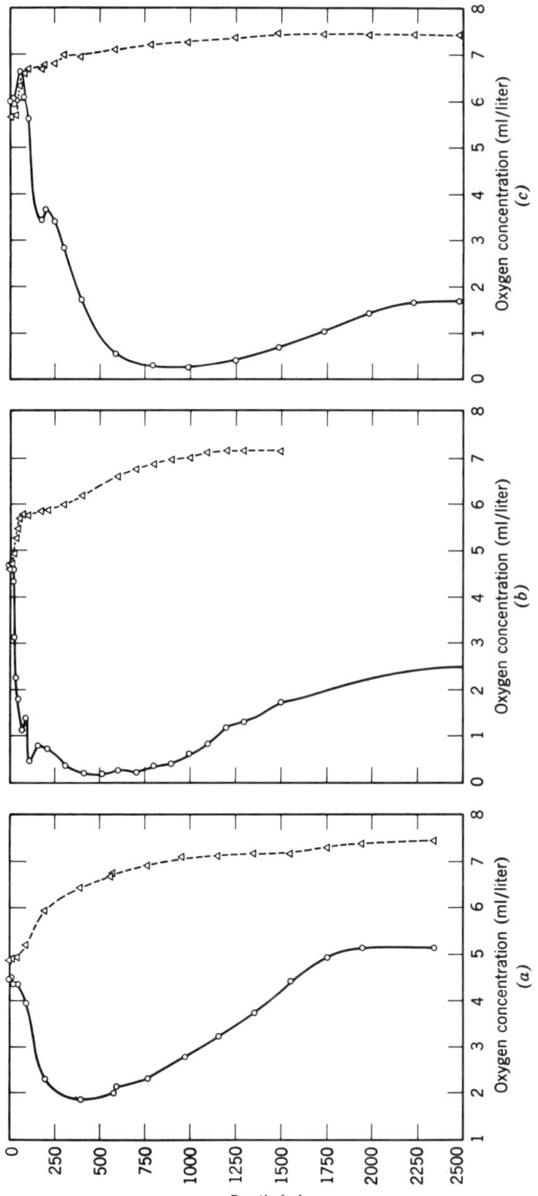

Fig. 6.1. Vertical distributions of dissolved oxygen, o, and of the amount of oxygen to be expected from the equilibrium solution of air, Δ. (a) A station in the tropical North Atlantic Ocean. (b) An eastern tropical North Pacific Ocean station. (c) An eastern North Pacific Ocean station. The area between the two curves is an approximate measure of the amount of oxygen consumed from the water. (From John F. Brahtz, *Ocean Engineering*, New York: Wiley, 1968.)

impart to sea water its *buffer capacity* or resistance to change the pH. The carbonate system in the ocean is part of one of wider geochemical significance in which carbon is participating in an interacting cycle between the biosphere, the lithosphere, the atmosphere, and the oceans.

Sea water is slightly basic, having a pH of 8.0 to 8.2 at the surface and decreasing to values around 7.7 to 7.8 with increasing depth. This alkalinity of excess cations arises principally from the carbonate, bicarbonate, and borate ions. The pH, the alkalinity, and the solubility of calcium carbonate are closely related in sea water. The buffer capacity is of great significance in natural bodies of water where the pH tends to be changed by the addition or removal of carbon dioxide principally by respiration and photosynthesis, respectively. The pH is also affected by temperature and pressure. An increase in these environmental elements decreases the pH. Pressure also affects the solubility of calcium carbonate and this, together with pH behavior, is responsible for important precipitation processes in the marine environment.

It can be concluded that the dissolved gases impart certain characteristics to sea water. The possibility of sustaining life higher than bacteria depends on the concentration of dissolved oxygen. Low concentration of dissolved oxygen may affect the presence of marine organisms, but it is very unlikely that variations of nitrogen and argon have any sensible biological affect. From the viewpoint of waste disposal, oxygen is instrumental to the efficient biochemical decomposition of such wastes. Decomposition of organic matter can take place in the absence of dissolved oxygen, but the products are toxic. Primarily because of the carbonate system, sea water is a buffer solution; that is, it resists changes in pH and also maintains the ratio of carbonate to bicarbonate ions. This may have significant biological implications.

BIOCHEMICAL CYCLE

The composition of sea water can be altered by boundary exchanges of matter and energy or by in situ processes that take place within the body of water itself. The first mode of exchange includes the addition or removal of water through the sea surface by processes of evaporation, precipitation, runoff, and the freezing and melting of the sea ice, which not only changes the salinity but the ratio of the major ions. Solutes can also be exchanged across the sea surface by leaching of the land, precipitation, and sea spray. They can also be removed from the water by the organisms for growth and deposited as sediment when they die. The extensive deposits of globigerina and diatomea oozes are typical examples of the removal of calcium and silica by the action of the plankton. Solids can be added to the ocean by agents, such as river-water sediment load, dust blown by the wind, subaerial and submarine volcanoes, among others. These terrigenous materials are, in most cases, subjected to other processes in the marine environment that, when deposited in the sea floor, sediments may have been altered considerably.

The in situ changes are produced by diffusion, advection, and biological activity. The latter is of particular importance because the life processes take place rapidly in relation to the life of the water mass and their study can help considerably in the physical description of the sea. For this reason, and because of its ecological implications, the biochemical cycle will be discussed in detail.

The influence of organisms on the composition of sea water is conditioned by physiological influences and shows the regularity inherent in organic processes. Elements are extracted from sea water by the primary food producers—the phytoplankton—in the proportions required to produce organic matter and are returned to it as excretions and decomposition products. However, in studying the changes produced in the composition of sea water by the marine organisms, a distinction must be made between a conservative and a nonconservative element or property. An element or property whose concentration within the water mass is unaffected by biological activity is conservative. Typical conservative concentrations are the heat content (temperature) and the major solutes—salinity. Dissolved oxygen and many of the minor constituents of sea water, particularly the nutrient salts, are strongly affected by biological activity.

The conservative concentrations are transported from place to place by advection and move from one parcel of water to another by eddy diffusion. The nonconservative elements are affected by the same physical agents but they also may move from one water layer to another in additional ways, such as by the vertical displacements of organisms and by the sinking of organized matter under the force of gravity. These movements of organic matter combined with the selective absorption of certain elements by organisms provide a mechanism for the fractionation of the components of sea water and the redistribution of these elements in characteristic patterns. This type of biochemical circulation explains the difference in the distribution of conservative and nonconservative properties of sea water.

The proportions in which the elements of sea water enter into the biochemical cycle are conditioned by the elementary composition of the plankton which constitutes the bulk of the biomass. The principal elements present in plankton are carbon, nitrogen, and phosphorus. The average ratios are 106:16:1 by atoms or 40:7:1 by weight. The variation of the inorganic forms of these elements in the ocean takes place in the same ratios which support the concept of their biological utilization. The individual cycles of nitrogen, phosphorus, and carbon have been determined in the different oceans. Silicates have also been studied in detail because they are required for and consumed by the growth of diatoms. To understand their distributions and cycles, it is convenient to describe in a general way the mechanisms of the biochemical circulation.

In the upper layers of the ocean—the euphotic zone—which light penetrates, phytoplankton activity forms new plant material. This material is made up from substances in solution in water—the carbonate components and nitrogen and phosphorous compounds. They are removed in the proportions stated and their transformation into plant tissue constitutes the first link in the food chain of the sea. This process is commonly referred to as *primary productivity* or *first trophic level* of the food web in the sea. The next stage involves the zooplankton eating the phytoplankton, which in turn are eaten by the carnivorous, and so on. Each transformation constitutes a higher trophic level. During these *particulate* stages, when the nutrients constitute the living or dead solid matter, they can be moved independently of the motion of the water itself by swimming and gravity and they are distributed in a different pattern than the major constituents, which are conservative and are distributed essentially identically. The net motion of the particulate organic matter is downward and below the euphotic zone—where light is not available—the net biological process is respiration, by which matter is returned to the dissolved inorganic form (remineralization). These processes constitute a permanent loss of nutrients from the upper layers, and only the vertical motion of the water can return the remineralized nutrients to the euphotic zone—where photosynthesis can proceed and the cycle can begin again. However, the refertilization of the upper lighted layers by the vertical motion of the water has a seasonal cycle because of environmental implications. This is particularly true in lakes and temperate latitudes where the density stratification plays the key role. During the summer, the insolation of the sea surface creates the strong seasonal thermocline which results in a density stratification that reduces or prevents vertical mixing. The growth and sinking of the phytoplankton leaves the surface layers devoid of nutrients, which are transported to the dark deeper waters. The depletion of nutrients and the reduction of insolation are responsible for very little photosynthesis at the end of summer. During the fall, the cooling of the upper layers creates negative stratification, and sinking of the surface takes place. The increase of winds in winter helps this vertical mixing of the waters and the nutrients are brought to the surface from the enriched deeper layers. The nutrients will then be almost uniformly distributed vertically, and conditions are ripe for photosynthesis to begin as soon as light is available in appropriate amounts during spring.

This is the typical seasonal cycle in the temperate latitudes with annual refertilization of the upper layers by deep mixing of the water column in winter. In other parts of the ocean, certain dynamic features may prevent or enhance the resupply of nutrients to the surface. The central oceanic gyres, of which the Sargasso Sea is an example, are permanently stably stratified, and vertical motion of the water is prevented. The surface layers are practically devoid of nutrients and primary productivity is very low. The contrary is true where upwelling takes place. This condition of continuous or nearly continuous nutrient supply from the deeper water is found generally along the west coast of continents where prevailing winds create this particular circulation and in regions where topographic submarine features or variations in the advective oceanic circulation induce upward motion of the water.

Typical vertical distributions of oxygen and nutrients are shown in Figs. 6.1 and 6.2, respectively. In general, it is found that the surface layers are saturated with oxygen from the air. In the homogeneous, wind-mixed layer explained in Chapter 3, the dissolved oxygen and nutrients are uniformly distributed. Under strong stable stratifications near the surface, large vertical gradients in the oxygen and nutrients develop together with very small concentrations of the nutrients at the surface layers and a subsurface oxygen maximum of biological origin. This maximum is found in the upper 50 m with concentrations almost twice that corresponding to saturation with air. Below this maximum, the dissolved oxygen generally decreases and the nutrients generally increase with increasing depth. The nutrient maximum is found at 1000 m, and the oxygen minimum at 600 m. At greater depths, the oxygen again increases and the nutrients decrease, but these changes are related to the replenishment of the deep water by the thermohaline circulation explained in Chapter 4. The distributions of the dissolved oxygen and of the nonconservative nutrients support the concepts expressed in the biochemical circulation.

PRODUCTIVITY OF THE OCEANS

The chemistry of marine food production is as far ranging as the science of biochemistry itself. Even the consideration of the processes involved in the initial stage in the food web in the ocean, the production of sea plants, will be a tremendous task. However, the hope that the oceans will overcome the shortage of animal proteins on land for the ever-increasing population has considerably increased the efforts in trying to understand the quantitative processes involved in the food chain in the sea.

The primary producers in the marine environment is the group of microscopic organisms known collectively as the *phytoplankton.* By far, they surpass the large attached (benthic)

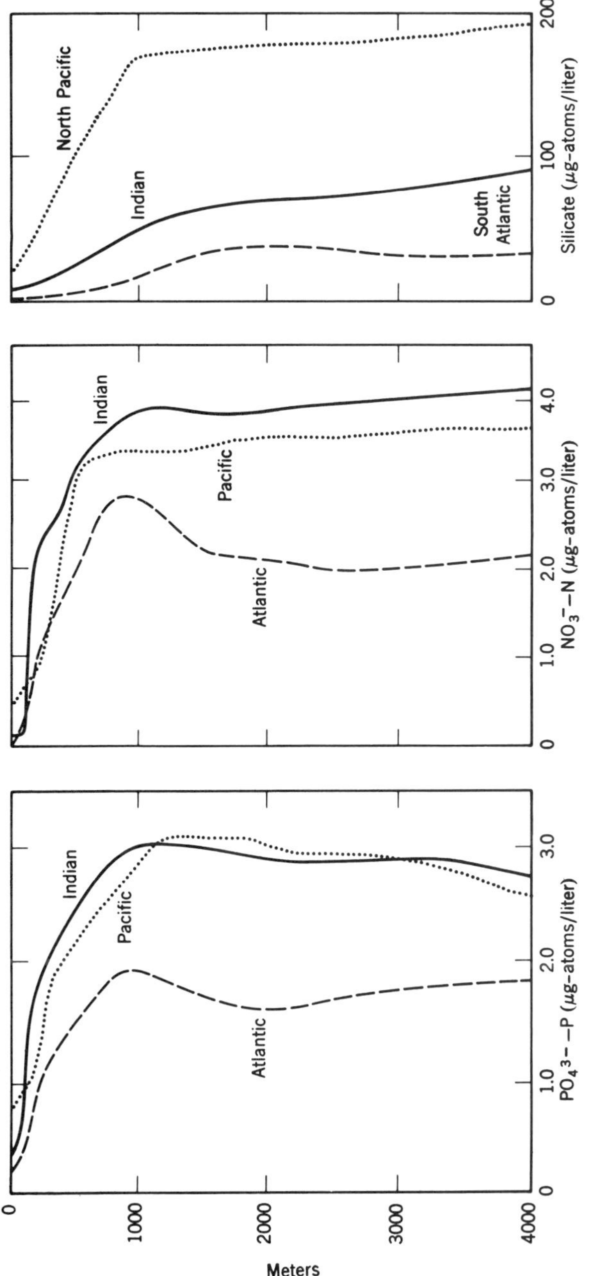

Fig. 6.2. Typical vertical distributions of nutrients. (Adapted from Sverdrup et al., 1942, by L. K. Coachman.)

algae found in shallow waters. Photosynthetic plant production in the ocean is of paramount importance because it initiates the food web in the sea, which terminates in the larger fishes and sea mammals. Included under the general term *plankton* are the organisms suspended in water, without mobility or with very little mobility, which are at the mercy of the water movements. The division in *phytoplankton* and *zooplankton* is to separate the plant and animal, respectively.

To carry on the process of photosynthesis, the unicellular marine phytoplankton depends on three environmental variables. They are visible light (intensity and duration of illumination), temperature, and the chemical composition of the environment. The effect of these parameters on the growth of phytoplankton is one of the main problems in biological oceanography, and it must be solved if the production of plant food in the sea is to be understood.

When discussing the fertility of the sea in terms of oceanic production, it is important to make clear what is meant by the different expressions used. The word *production* has very often been used as synonymous with *standing stock;* this severe confusion should be avoided. By standing stock, we mean the quantity of autotrophic plants at a given time. It is synonymous with standing crop. However, the term *standing stock* is also used for the quantity of organisms at other trophic levels; for example, herbivorous zooplankton organisms are standing stock.

Since the term production is a broad one, expressions such as *gross primary production* and *net primary production* have been introduced. These definitions are related to the method of measurement used. Gross primary production is equivalent to the rate of real photosynthesis, and net primary production is this value minus the rate of respiration by the algae. All these quantities are given as carbon fixed (or released) per square meter or per cubic meter per unit of time, which is usually 24 hr.

There are different methods for measuring primary production. The rate may be measured either directly or indirectly by estimating the standing stock of phytoplankton and using a conversion factor. All these methods have their advantages and disadvantages and all have contributed to the present knowledge of marine production. Finally, it is even possible to estimate theoretically the relative size of the primary production in oceanic areas where the hydrographic conditions are known. It is particularly important to know the degree of the intermixing of the euphotic surface layers with nutrient-rich subsurface layers. Present knowledge of primary production in the ocean has shown a remarkable correlation with oceanographic conditions. On this basis and with some limitations, the rate of organic production may be treated in the same way as physical or chemical properties, such as temperature and salinity.

Information of primary production is basic to evaluate the role of the ocean as a food provider, but it must be kept in mind that the marine food chain differs from that on land in that there are generally several steps between the plants and the larger invertebrates. A necessary condition for a large-scale fishery in a certain oceanic area is that a significant part of the primary production is within the area, but this may not be the only necessary condition. It is essential to investigate the transfer of organic matter between the different trophic levels; the present knowledge is still far from desirable.

The complexity of the food chain in the marine environment is suggested by Fig. 6.3, which indicates the state of affairs in a hypothetical (large) body of water in the euphotic zone in a productive region of the ocean. The area of the blocks is proportional to the standing stocks of the components and the width of the arrows showing the ways that energy and mass are transferred are proportional to flow rates.

Because of the ever-increasing importance of extracting proteins from the sea, the search for biologically active substances in sea water is the modern attack of the old problem of *bad* and *good* waters. It is a clear fact that water masses differ in capability to sustain growth of various organisms, and the expressions *good* or *bad* waters refer to that capability. In this sense, the most productive or fertile areas of the marine environment are near the coasts, above shallow bottoms (banks), and in regions of water mixing (merging of different bodies of water, upwelling, etc.). Because all life in the ocean depends on the phytoplankton as primary producers, and because there exist simple techniques for measuring primary productivity aboard research vessels, the available information of the

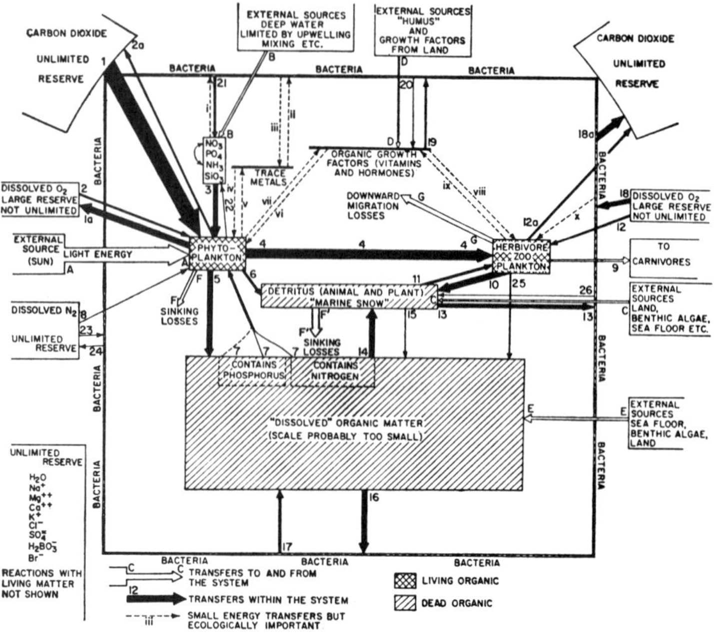

Fig. 6.3.   Diagram showing the idealized food web in the marine environment in the early stages. (From J. P. Ryley and G. Skirrow, *Chemical Oceanography*, New York: Academic Press, 1965.)

oceanic regions with high rates of photosynthesis has increased considerably in the last 15 yr.

The characteristics and behavior of the chemical and organic marine environment are of great importance to the oceanographic engineer. The presence of so many chemicals dissolved in sea water and the activities of the marine organisms affecting not only the chemical system but also the physical behavior of the waters of the ocean presents challenges of different intensity for the engineer working at sea. Some of these challenges to be met are the corrosive nature of sea water and the damaging activities of the marine fouling and boring organisms. Moreover, the proper use of the sea as a natural dumping ground for different wastes, such as domestic sewage, industrial wastes, radioactive wastes, heat, agricultural drainage, and brines from saline water conversion or chemical recovery plants, among others, requires a thorough understanding of the chemical and biological behavior of the marine environment.

As for the sea as a source of commercial products, including fisheries, there is a growing body of literature presently available.

SUMMARY

The marine environment is a complex solution of many salts whose concentrations are changing at the air–sea boundary by interactions with the atmosphere and within the water mass itself by biological processes, particularly photosynthesis in the euphotic layer and remineralization of organic matter at greater depths. Greater changes take place where rivers discharge their waters into the ocean. Gases are continually being interchanged with the atmosphere at the air–sea interface where saturation conditions for the dissolved gases in the water can be assumed as a normal state. The vertical distribution of the biologically affected oxygen and carbon dioxide gases reflect the photosynthetic and respiratory processes in the euphotic layers and the remineralization of organic matter below this zone. The vertical distribution of the nutrient salts, particularly nitrogen compounds, phosphates, and carbon, is also affected by these processes.

The food web in the sea is more detailed than its counterpart on land—there are more trophic levels. The beginning of the food chain in the marine environment is the photosynthetic activity of the phytoplankton—primary production—and the second link or trophic level is the zooplankton feeding on the phytoplankton.

Primary production is higher in the fertile areas of the ocean where a combination of the oceanographic physical processes and the biochemical system are responsible for such a fertility.

The chemical and biological nature of the marine environment have the following principal implications on the activities of the ocean engineers:

1. The ocean is a highly corrosive medium.

2. Biological activity produced by the lower organisms that attach to and bore through man-made platforms. The larger animals bite the ropes that anchor buoys and hold instruments.

3. Biological noise made by marine life affects the performance of acoustic devices.

4. The sea is a major source of food, minerals, and energy. The techniques for a sound exploitation of these resources present a real challenge to the engineer.

7

The Earth Beneath the Sea

It has been stated that the sea bottom constitutes the lower and lateral boundaries of the world ocean. This benthic boundary, as most of the interfaces in geophysics, is endowed with a variety of complex phenomena, and its understanding is essential to solve many scientific and engineering problems.

The sea bottom and the rocks beneath it contain valuable information about the history of our planet. The study of the physical nature, mineralogy, and fossil biota of the sediments that carpet the ocean floor provides the tools to understand present and past oceanic conditions including the processes affecting erosion and deposition, rates of sedimentation, sources and character of materials constituting the sediments, past climates, and eustatic changes in sea level. It also assists in the understanding of the geochemical cycles and processes of organic evolution as well as secular variations in the patterns of oceanic and atmospheric circulation, sea water temperature fluctuations, and permanency of continents and ocean basins.

The sea bottom consists of a rough rock topography which is covered by the marine sediments and, in many places, outcrops through them. In studying them, it will be possible to understand many first-order geological problems, such as the history of ocean basins, crustal uplift and subsidence, areas and periods of volcanism, the nature of the spatial and temporal variation in the composition of melts from the mantle, the age of the oceanic crust, the extent of polar wandering, and the possibility of continental drift.

However, before becoming involved in the description of the sea bottom, it is appropriate to discuss some basic geological

135

differences between continental blocks and ocean floor. The earth's crust in the continents is generally composed of material less dense than the material making the oceanic crust. The land rocks are rich in aluminum silicates, whereas the denser rocks underlying the ocean contain more iron and magnesium. The fact that the land is lighter and stands higher than the sea floor and that this difference is not reflected in the gravitational field over both environments suggests that the crust of the earth is floating on a still denser mantle. The zone of separation between the crust and the mantle is identified by a sharp discontinuity in the velocity of propagation of earthquake waves, which occur under both the continental and oceanic crustal rocks. This transition zone is known as *Mohorovicić discontinuity*. Seismic data taken to determine the depth of this discontinuity have shown that the thickness of the crust under the continents is on the average 33 km, whereas the crust under the ocean is much thinner and averages 5 km. This and the difference in density of continental and oceanic crustal rock appear to indicate that they are different material. The transition between the continental crust to the oceanic crust takes place gradually at the outer end of the continental margin. Figure 7.1 illustrates the geological differences.

Knowledge of the continental margins and deep sea floor is important to most of the oceanographic disciplines: to physical oceanography because the topographic features create oscillating basins with their natural periods of oscillations interacting between them, because it affects the free flow of water and introduces special characteristics in the pattern of circulation and in the characteristics of the different water masses, and because a great deal of the kinetic energy of the water masses is dissipated through friction in the ocean floor; to biological oceanography, because the presence of land barriers or submarine ridges affects the distribution of organisms and of the ecological relationships; to oceanographic engineering because the knowledge of the sea bottom relief, mechanical properties of ocean sediments, erosion and deposition by waves and currents, and mass flows on the sea floor are essential in the proper design of marine structures on or near the sea bed; and to marine law to aid in the solution of the many difficult legal problems.

The benthic boundary is not a solid boundary in the full

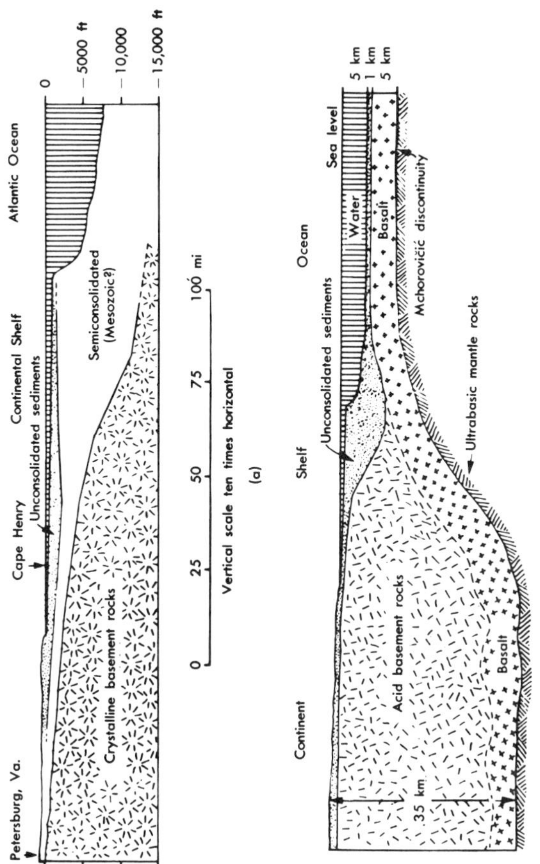

Fig. 7.1. (a) Section across the Atlantic coastal plain, as measured by Ewing. (b) Generalized section across a stable continental margin. (Reprinted from *The Earth as a Planet*, 1954, G. P. Kuiper, editor, by permission of the University of Chicago Press.)

sense of the word, but a transition from fluid to fluid with suspended particles, to solid with interstitial fluid, to solid. The characteristics of this boundary are poorly known and its detailed nature is the central topic of submarine geology. The knowledge of this important interface is still far from adequate despite the considerable increase in the capabilities for study during the last 15 yr. Development of new instrumentation, equipment, and techniques for positioning the ships at sea, together with the increasing number of research vessels engaged in oceanographic work, has given considerable impetus to the study of submarine geology and geophysics; and if this trend continues, it will enhance considerably the knowledge of the oceans.

A major advance in the field has been the physiographic outlook of the submarine topography replacing the classic bathymetric representation. The common nautical charts showing the isobaths with special emphasis in topographic features, which may constitute a danger to surface and submarine navigation, and the *General Bathymetric Charts of the Oceans* at a scale 1:10 million serve very useful purposes for many ocean activities. However, they are poorly suited for investigations of sea bottom physiography because the isobaths typical of the bathymetric system cut across physiographic units with no particular interest in the physiographic boundaries determined by a careful study of the sea floor relief. The main difference between the physiographic and the topographic approaches lies in the genetic side of physiography, which impresses upon it a more scientific interest. However, this does not imply that physiographic classifications need not be purely genetic ones. Actually, the present knowledge of the sea floor relief is not adequate to favor the use of an entirely genetic classification.

Another area of knowledge where considerable advances have been made is the one related to the distribution and nature of the sediments. Responsible for this progress have been the improved instrumentation in seismic techniques and the increasing number of sediment cores collected from the deep-sea floor. This basic field information has given good material to the stratigrapher and paleoceanographer for their research in elucidating the history of the oceans and the earth. The present capability of drilling into the deep-sea floor opens a new avenue to the progress of those fields.

SUBMARINE RELIEF

The topography of the planet Earth shows the presence of two levels which comprise most of the earth's surface: the one a few hundred meters above sea-level, which represents the normal surface of the continental blocks; the other far below sea level between 4000 and 5000 m, which comprises over 50 percent of the surface of the earth and represents the deep-sea floor. The sea bottom can be divided into three major morphologic divisions: *continental margin, ocean-basin floor,* and *midoceanic ridge.* These major divisions admit subdivision into categories of provinces which can also be divided into individual physiographic provinces.

Continental Margin The continental margin is the transition zone between the continent and the ocean floor and is generally made up of three provinces: *continental shelf, continental slope,* and *continental rise.* For practical reasons, the continental slope and continental shelf are often referred to as *continental terrace.* The continental terrace includes the greatest topographic feature on the surface of the earth, an escarpment $3\frac{1}{2}$ km high extending over 350,000 km in length, which is in turn the visible manifestation of the greatest structural discontinuity on the earth's surface—the transition from continental to oceanic crust. A typical profile across the continental terrace off the northeastern United States is shown in Fig. 7.2.

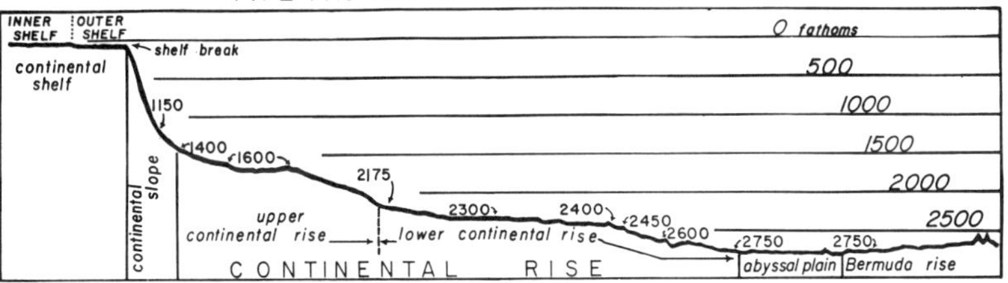

Fig. 7.2. Profile plotted from P.D.R. records. This profile is representative of the sector from Georges Banks to Cape Hatteras, eastern United States (after Heezen et al., 1959). (From M. N. Hill, *The Sea*, Vol. 3, New York: Wiley-Interscience, 1963.)

The structure of the continental margin is related to the formation, permanency, and drift of the continents. The sediments deposited in the continental terrace are of special geological and statigraphic importance because they help in the interpretation of sediments found in land and probably deposited in similar environments. By different agents, these sediments are transported to the ocean basins and play an important role in the sedimentary layer found in the deep ocean. In the continental margin, a coastal interface is also found with its rich and complex phenomena known as *nearshore processes*, which include those where beaches are formed and disappear and those pertinent to the origin and role of *submarine canyons* that cut deeply across the shelf and extend to the continental slope.

The continental shelf is generally defined as the natural continuation of the continents into the sea down to 100 fathoms or 200 m. At around this depth, the *shelf break* is found and the continental slope begins and extends down to the abyssal depths with an increase in declivity. If the 2000-m isobath is taken as a conventional outer limit of the continental slope, both the shelf and slope cover 7.6 and 8.5 percent of the sea bottom, respectively.

The definitions and limits of these features are not rigid for several reasons. First, the continental break is not always found at the same depth. The Antarctic continental shelf extends normally down to 300 fathoms (540 m), while in the Arctic Ocean, particularly in the east Chukchi Sea, it is found in only 35-fathoms (63 m) depth. Other shelves have several breaks or steps; and when the distances between steps are great, it is difficult to decide which one of the steps marks the end of the shelf. It is also found that some shelves are not connected to the continents, although they are not far from them, as is the case with the Bahama Banks. These are very shallow features and are surrounded by steep escarpments and deep water. They are known as *continental borderlands*.

Detailed surveys of the continental shelves have shown that the old concept of a flat, featureless plain is not a common occurrence. The increased use of the subbottom continuous profiler has provided very valuable information about the topography of the shelf under the veneer of sediments. The most fascinating features are, perhaps, the submarine canyons or valleys. Some have been completely covered by sediments and are difficult to find, while others, because of the presence of scouring agents, are kept sediment free. Another type of complicated topography is the shelf province along the coast of southern California known as the *basin and range* type. In this region, the real continental shelf is constituted of a narrow strip 2 to 25 km wide. Outside this strip and before reaching the abyssal depths, a province made up of basins and ranges is found. Finally, after the outermost ridge is crossed, a typical continental slope is found.

The great percentage of continental shelves whose topography resembles the aerial continental landscape suggest that they belong to the same structural units and were shaped by the same processes. Typical examples are the glaciated, drowned valleys in Norway and Scotland, the California borderland, and the wide, flat platform off Argentina where the shelf relief is very similar to the continental morphology.

The shelf break, which is recognized as the zone where an abrupt change in slope takes place, marks the end of the continental shelf and the beginning of the continental slope. This change of slope within the continental margin has been attributed to processes of sedimentation or abrasion by waves at present or past lowered sea level (wave built and wave cut, respectively). The advances made on the knowledge of wave action on the continental shelves has resulted in the conclusion that the change in slope was produced by wave erosion at a time when sea level was lowered to within about 5 fathoms (8 m) of the break.

The main topographical features of the continental slope are the *submarine canyons*. They are deeply incised, V-shaped valleys with steep walls, many rock outcrops, winding courses with many tributaries, and virtually continuous outward-sloping floors. These are known as the true submarine canyons as there are also other types of submarine valleys which extend down the continental slopes. These deep gorges attain varying dimensions: widths from less than 1 mile to 10 miles; depths from a few meters to several hundreds of meters; and lengths from a few miles to a few hundred miles. Canyons have been found along the coasts of many parts of the world and very few have been described in detail from their heads down to their termination on the outer fans. The increasing use of skin-diving techniques for scientific exploration has made it possible to observe interesting phenomena taking place at the shallow canyon head, particularly off California. The best studied canyon is perhaps the one located off La Jolla, California, which has two main branches coming very close to the coast. The origin of submarine canyons has been the subject of considerable interest in the last 30 yr; and from the many hypotheses offered, two have remained as potentially valid: erosion by turbidity currents and subaerial erosion followed by downwarping. The first one is becoming the more accepted.

The continental slope is not a uniform province even if the canyons are excluded. There are diversified features, such as rugged topography, terraces, steep scarps, ridges, transverse spurs, and associated embayments. Such a variety of features suggests that the continental slope, as well as the shelf, must have various origins.

The *continental rise* lies at the base of the continental slope and has generally a smooth topography with irregularities which seldom range more than 20 m. Many of the characteristic features found in this province are accounted for in the action of turbidity currents. The transition between the continental rise and the abyssal plain is generally characterized by a fairly abrupt change in gradient from a value greater than 1:1000 on the continental rise to one smaller than such a value on the abyssal plain.

It is difficult and unlikely to be able to attribute to one particular process the origin and shape of the continental margin. From the ever-increasing amount of information that is becoming available, it is evident that although some tectonics have played a role in the earliest history of the continental terrace, alternate cycles of sedimentation and marine flattening on the continental shelf and long continued abrasion by slumps, turbidity currents, and deep oceanic currents on the continental slope together with a general sinking of the area could alone have produced the characteristic form of the continental margin.

Ocean Basin Floor   The next major morphologic division comprises the broad floors of the ocean basins. About one third of the Atlantic and Indian oceans and three fourths of the Pacific Ocean are included in this division. As was the case with the continental margin, the ocean basin floor can be divided into two provinces: (1) abyssal floor and (2) oceanic rises.

The topography of the abyssal floor is characterized by areas of flat bottom or abyssal plains, abyssal hills, and areas of undulating smooth topography. The abyssal plain is an area of flat bottom with a slope less than 1:1000 and varying between 1:1000 and 1:10,000. The transition of the continental rise to the abyssal plain takes place through a gradual merging of these two provinces often with the presence of abyssal cones and deep-sea fans, although in many areas, deep-sea channels from submarine canyons extend onto the abyssal plains. Abyssal hills and seamounts break through the surface of the plains in such an abrupt way that they do not disturb their smooth slope. An abyssal hill is a small, sharp hill that reaches a height of a few fathoms to a few hundred fathoms and a width of a few hundred feet to a few miles. In general, it is common to find the abyssal hills at the seaward edge of the abyssal plains. The areas of undulating topography may be the result of either smaller tectonic activity, that which formed the abyssal hills, or they may be older features and, consequently, have a greater thickness of sediment cover.

Many theories exist concerning the origin of the abyssal plains, such as areas of no tectonic activity, lava plains, sub-aerial erosion surfaces, and areas of long continued sedimentation, among others. The most modern one is the one which considers turbidity currents filling the original topography with sediments. This theory is appealing because it not only accounts for the existence of abyssal plains, but also explains many of the characteristic features present in the continental rise.

Some scientists define oceanic rises as "aseismic areas slightly elevated above the abyssal floor which do not form parts of the continental margin or the mid-oceanic ridge." Others use a broader, less restricted definition. In contrast to the midoceanic ridges, the oceanic rises have no seismic activity except where active volcanism is present; their relief is more subdued and they are asymmetrical in cross section; heat flow appears to be normal except near areas of active volcanism; gravity values are abnormal rather than subnormal and other geophysical differences are also exhibited. The best known examples of rises are the Bermuda Rise in the North Atlantic (Fig. 7.3) and the Hawaiian Ridge in the Pacific.

The ocean basin floor is generally aseismic. The small number of existing volcanoes and volcanic islands has developed the concept that the ocean basin has been tectonically stable in geological time. However, with the improvements in the

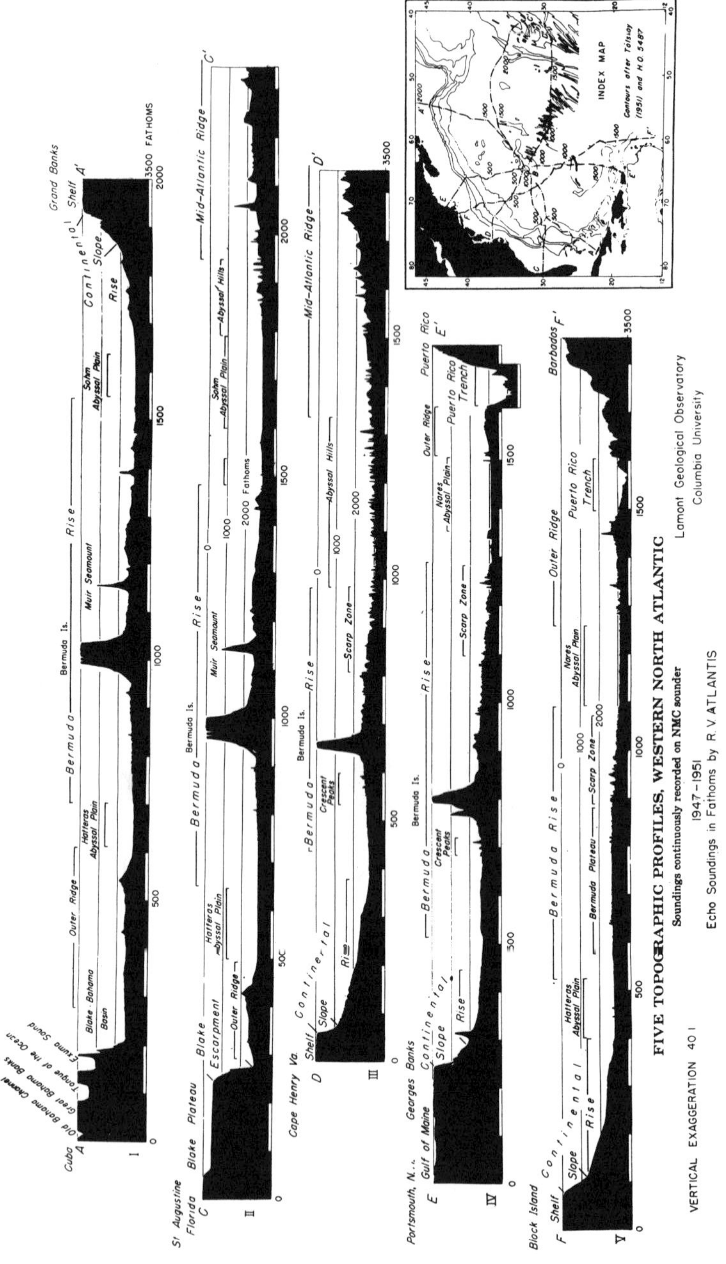

Fig. 7.3. (After Heezen et al., 1959.) (From M. N. Hill, *The Sea*, Vol. 3, New York: Wiley-Interscience, 1963.)

techniques for geologic and geophysical surveys, it has been found that the basin floor is covered with volcanoes and broken in many areas by enormous faults which means that it has not always been stable. The problem now is to determine whether the present is an abnormal quiet tectonic situation or whether volcanoes and faults can be properly produced in a relatively stable basin in geologic time. The answer to this question depends on how long back in time one can go through the dating of old sediments. The feasibility of drilling in the deep ocean has improved this situation and Upper Jurassic rocks (around 150 million yr old) have been collected.

Midoceanic Ridge The third major geomorphological division is the midoceanic ridge. It is the greatest topographic feature of the sea bottom and extends continuously through the Atlantic, Indian, Antarctic, and South Pacific oceans and has a length of over 30,000 miles. The width is generally over 1000 km and the height of the elevation with respect to the adjacent sea floor is 1–3 km. The axis of this global ridge system follows the median line of the ocean and its size is such that it covers the center third of the ocean. The crest of the ridge is characterized by (1) a rift valley of 900–2700 m deep and 15–30 miles wide, (2) the mountains which form the sides of the rift valley, and (3) the rugged plateau bordering the rift mountains. These provinces are found in the Atlantic and Indian oceans, but in the eastern Pacific the morphological aspect of the ridge is more subdued. The flank provinces on each side are characterized by steps separated by large scarps. The midoceanic ridge also shows seismic activity along the axis; measurement indicates both high and subnormal heat flows in the central region and a distinctive pattern of magnetic anomalies of similar characteristics on both sides of the ridge. The origin and structure of the midoceanic ridge have been the subject of much speculation by geologists and geophysicists. The present belief shared by many scientists is that it is the result of upwelling of mantle material. The distinctive geophysical relations, the location in the midocean, and the idea that the mirror-image pattern of the magnetic anomalies on the flank provinces represent periods of reversals in the earth's magnetic field and spreading of the sea floor all suggest upwelling of mantle material. Moreover, the age and thickness of sediments on both sides of the ridge add support to the theory of upwelling of mantle material and sea-floor spreading. Figure 7.3 shows the location of the midoceanic ridge in various sections across the North Atlantic.

There is another topographic feature of geophysical importance which cannot be included under continental margin or ocean basin floor. This feature constitutes the deepest part of the oceans, known as *trenches*. Trenches are present in the three oceans, particularly in the Pacific Ocean. The arcuate shape of these features, commonly associated with island arcs, mountain ranges, and volcanism, as well as a considerable body of data on gravity, crustal structure, heat flow, and earthquake activity strongly suggests that trenches are surface manifestations of large-scale processes taking place within the earth.

Trenches and island arcs were among the first topographic features of the ocean to be surveyed because of the impediments of laying submarine cables. At present, the greatest depth known is about 10,600 m in the Marianas Trench, although almost similar depths have been found on other trenches. The Puerto Rico Trench has been investigated in great detail, and precise computation of mass distributions were performed that fit observations of gravity, seismic velocities, and indicated thickness of layers. Moreover, it has been found that sediments were not disturbed and that the trench is far out of isostatic equilibrium. The suggested explanation about the forces necessary to produce the structure and departure from isostatic equilibrium has been attributed to mantle convection currents.

SEDIMENTS

The processes by which sediments are deposited on the sea floor, including the transporting agents which move the grains from one place to another and distribution and thickness, is the concern of marine sedimentation. What is known as statigraphic interpretation of sediments involves not only the geological study of the sedimentary rock but also the processes whereby their constituents are transported from the sources to the places of deposition.

Sediments found in the deep ocean come from several

sources, such as the products of continental erosion transported to the ocean by rivers, glaciers, and wind; shells and skeletons of microscopic organisms fall to the sea bottom after the death of the animals and plants that built them, volcanic materials from eruptions occurring both above and beneath the sea surface, elements and compounds dissolved in sea water which interact with other material and precipitate as mineral, and cosmic sources introducing extraterrestrial particles through the atmosphere.

The broad outline of present-day sedimentation in the open seas can be presented in a simple way. Off coasts with large rivers or glaciers, the dominant component of the deep-sea sediments is of terrigenous origin produced from the erosion of the continents; this occurs provided that the flow of sediments has not been prevented and trapped by ridges or trenches near the coasts. Depositional areas in the deep-sea floor, presently being fed with terrigenous detritus are the northeast Pacific, the northwest and southeast Atlantic and the Arctic and Antarctic oceans. Sediments transported by the wind are found everywhere in the sea bottom, particularly in midlatitudes leeward of continental deserts. Calcareous and siliceous sediments of organic origin occur in areas of great plankton population with a high rate of organic production. These areas are the equatorial regions, the subarctic and subantarctic belts, and near the western edges of continents. Near volcanic islands and seamounts, volcanic sediments are found and they constitute a major component of the deep-sea sediments, particularly in the western Pacific. In general, in areas such as the central, north, and south Pacific and the southeast Indian Ocean, which are far away from the major sources of solid materials, sediments accumulate more slowly.

The main transporting agent seems to be turbidity currents which dominate all other depositional processes in the deep sea. In many regions, sediments transported by these currents are very well known. The explanation of how this current generates and spreads is as follows: sediments collect in unstable masses in the head of the submarine canyons or on the continental slope where, through the triggering action of some agent, it slumps down and moves through the canyon or down the slope. Upon reaching the deep sea, a full turbidity current is capable of moving for hundreds of kilometers across a very gentle slope. Research on turbidity currents is conducted through laboratory experiments, studies of submarine canyons, measurements of the distribution of materials in the deep-sea fans near the canyon mouths, and observations from deep submersibles of the detailed morphology of the submarine canyons.

The possibility of high-density, high-velocity turbidity currents was proved by a study aimed to explain the breakage of submarine cables and the origin of deep sea sands. It was concluded that "events associated with the Grand Banks earthquake of 1929, November 18 may be considered as a full-scale experiment in erosion, transportation, and deposition of marine sediments by a turbidity current in which the submarine telegraph cables served to measure its progress and to give tangible evidence of its force. . . . " According to the explanation given by the authors, each successive cable was broken in sequence by a turbidity current which had originated as a series of slumps in the epicentral area. The current traveled down and across the continental slope, continental rise, and on to the abyssal plain, spreading as far as 450 miles from its source area on the continental slope. The velocity reached by the current in the continental slope exceeded 50 knots and on the abyssal plain was still over 12 knots. Figure 7.4 illustrates

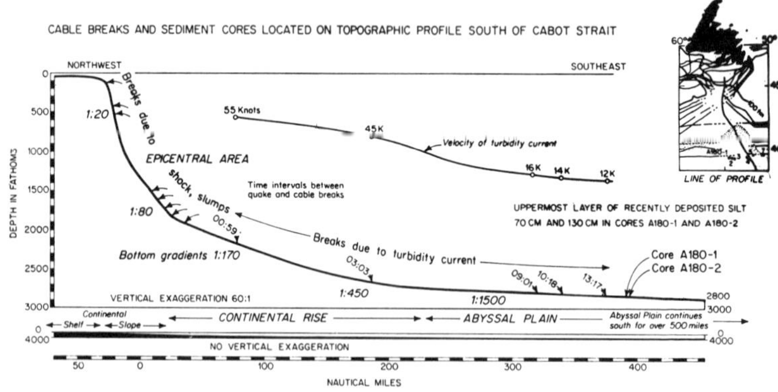

Fig. 7.4.   The superimposed graph is of the velocity of the Grand Banks turbidity current (after Heezen and Ewing, 1952). From M. N. Hill, *The Sea*, Vol. III, New York, Wiley-Interscience, 1963.)

this case. Studies of turbidity-current sediments deposited in the abyssal plains of the North Atlantic presented interesting evidence of turbidity-current beds consisting of silt and sand and sometimes even gravel.

The requirements of turbidity currents are sediment, a trigger mechanism, and a slope. There is some evidence for several different trigger mechanisms, such as earthquakes, hurricanes impinging on the shore, high bedload discharge of rivers, and even by slope failure resulting from gradual oversteepening of a depositional slope. This evidence came mostly from the study of cable failures from various parts of the world.

Other important agents responsible for eroding and transporting sediments in the deep sea are the bottom currents. Evidence of the sedimentation processes related to bottom currents is abundant. Ripple marks and scour marks shown in deep-sea floor photographs are simply astounding. Measurements of deep-sea currents have shown velocities up to 43 cm per sec (0.83 knot) at a depth of 4000 m and that these currents vary in direction and velocity over very short periods of time. Such strong currents explain the occurrence of ripple marks, scour marks, and denuded surfaces in the deep sea.

There is now little doubt that submarine currents of water and mud shape the ocean floor by erosion, solution, scouring, and transporting the sediments. Deposition is as important a process in the deep sea as erosion is on land.

Knowledge of the physical properties of marine sediments is essential to rational design of construction on the sea floor as well as to the interpretation of different types of sediments from geophysical techniques. Present knowledge of mechanical properties of ocean sediments has come mostly from submarine geological description and geochemical and biogeographical examination. The uppermost part of the recoverable sedimentary column may be studied in detail. Typical physical properties, such as elastic wave velocities, attenuation, density, and thermal conductivity are directly measurable. The bulk properties depend ultimately on the values of other observables such as water content, particle density, size and shape of particles, age, and chemical composition.

Considerable work has been carried out to determine the physical properties of sediments. The motivations have been of a scientific and practical nature. For engineering design, some of the descriptions have to be analogous to the refined measurements of terrestrial materials. It is difficult to measure specific volume, permeability, shear modulus and shear strength, compressibility, viscosity and thixotropy, erodibility, and possible gravity-elastic wave motion of deep-sea sediment brought to the surface because of the changes suffered by the internal structure of the material by the pressure changes. For scientific purposes, the physical constants most needed for interpretation of seismic data to provide information on sedimentary structure are dynamic values and values determined by relatively large-scale measurement. Because of the difficulties related to bringing the sample to the surface and the requirement of large-scale measurements, in situ determination of structural characteristics of bottom sediments is recommended.

LITTORAL PROCESSES

The *littoral zone* is the interface or transition zone between land and sea. It comprises about 8 percent of the surface of the world ocean, and about 30 percent of the world's human beings live near it. It is a complicated area rich in marine phenomena, with intense interactions among waves, tides, currents, and sediments. It is in this area that most of the wave energy of the ocean is dissipated, where the mixing and transportation of sediments begins, and where the beaches are formed and destroyed. The disturbing action of the shallow depths makes nonlinear the action of the waves and tides, which are the principal sources of inshore energy, and complicates considerably the interactions between fluid and sediment motion.

The transport of sand in the littoral zone is surprisingly large. Research in many coastlines of the world shows that the sand transport amounts to 10^6 cu m per yr. The main source of this volume comes from land by river discharge; only a small fraction is the result of cliff erosion. Some of the sand is lost from the littoral drift by various processes. Along coasts with broad continental shelves, the sand forms spits across embayments and there is a loss of sand by wind deflation. Along many steep coasts, notably off the west coast of the United States, the deep canyons extend up to the beach. These canyons intercept the littoral drift of sand which is diverted with its load of sediment into the adjacent deep basin.

The sediment is set in motion and transported by the energy that comes mostly from the dissipation of the wind-generated surface waves along the shores. This dissipation may take the form of different processes, such as reflection of wave energy from the beach or conversion to currents and fluctuations of other dimensions such as seiches, edgewaves, or turbulence. It may be transformed into heat or sound or used to move sediment. This sand in suspension is transported by what is known as the nearshore circulation system and deposited or removed from the beaches which are never at rest. A beach responds with great sensitivity to the forces that act upon it—waves, currents, and winds.

The central subject in the field of nearshore processes is the understanding of sediment motion, transport, and deposition. This requires an increased effort in the following areas: the mechanics of wave motion in and near the surf zone, the budget of water flowing in the nearshore circulation system, and the modes and mechanics of sand transport along the beach. In addition, it is necessary to increase the understanding of the extent to which models are capable of reproducing natural phenomena.

Before discussing the littoral processes in detail, it is convenient to define the limits of the different zones of the coastal environment. The littoral zone may be divided into three marine zones:

1. The *offshore zone* extending from the isobath where the waves begin to break down to the depth where the bed sediment admits disturbance.

2. The *surf zone* area of the sea bottom where the waves dissipate their energy by surface turbulence.

3. The *swash zone* area of the beach in which the residual wave motion consists of successive surges up and down the beach face.

It has been said that wave dissipation is the main source of energy in the sediment dynamics. The most important agent in shaping a shoreline and determining the characteristics of beaches is the intensity of the wave attack. Beaches, which are the sandy portions of a shoreline, respond immediately to changing wave conditions and may undergo changes in a matter of a few hours. Beaches, where the wave climate is highly variable, are more complex in both plan and profile than beaches which are subjected to less variable wave spectrum. The information on the wave climate is basic to an appraisal of the nearshore processes. These wind waves have been generated in the open sea by transfer of wind energy to sea surface water, which also generates surface wind currents. The swell propagates with very little loss of energy from the deep sea to the shallow coastal environment. While moving in deep water, there is a net small volume transport of water in the direction of their travel. This transport increases as the waves enter the shallow waters near the shore. The beach, however, presents a solid obstacle that reduces to zero the onshore momentum and energy is converted into a rise in water level on the beach. If, on the other hand, the direction of the wave propagation is not normal to the shore, a *longshore current* will develop. For reasons not yet well understood, this longshore current is confined to the surf zone. The breaking waves keep discharging water in the surf zone and feeding continuously the longshore current, and unless some compensation mechanism develops, the velocity of the current will go on increasing indefinitely with distance along the shore. The compensating device is the existence of outward-flowing *rip currents* at discrete points.

The net onshore transport of water by the waves in the breaker zone, the lateral transport along the surf zone by longshore currents, the seaward return of flow through the surf zone by rip currents, and the longshore movement in the expanding head of the rip current all constitute a typical nearshore circulation system (Fig. 7.5). The location and intensity of the rip currents are dependent on the submarine topography and configuration of the coast and the height and period of the waves. There is, however, some degree of variability in the nearshore circulation system. This variability is created by the irregular nature of the train of incoming waves in which a grouping of high waves is followed by low waves giving rise to a pulsating phenomenon known as *surf beats*.

Field measurements show that the water has a slow flow near the bottom in the surf zone between the rip currents which do

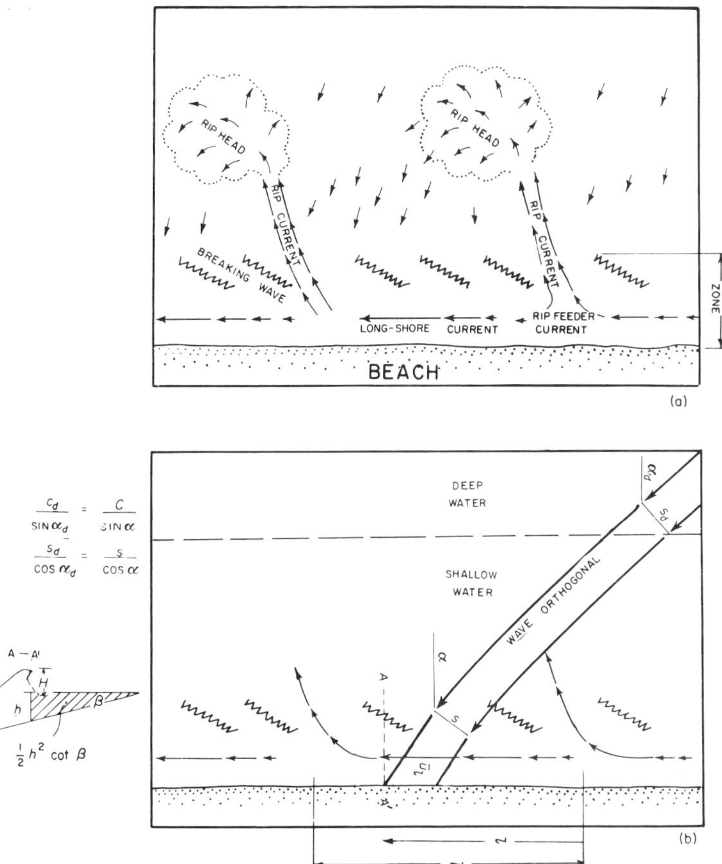

$$\frac{c_d}{\sin \alpha_d} = \frac{c}{\sin \alpha}$$

$$\frac{s_d}{\cos \alpha_d} = \frac{s}{\cos \alpha}$$

$$\tfrac{1}{2} h^2 \cot \beta$$

Fig. 7.5. Schematic diagrams of (a) the surface flow in the near-shore circulation system associated with wave action in and near the surf zone (after Shepard and Inman, 1951) and (b) wave and current relations in the proposed model for longshore transport of sand. (From M. N. Hill, *The Sea*, Vol. 3, New York: Wiley-Interscience, 1963.)

not extend beyond the surf zone. Water outside the breaker zone moves toward the beach. Bottom topography influences considerably the longshore currents due to the differential refraction of the waves approaching the beach. In such a case, the intensity of the current depends on the gradient of the breaker height along the beach as well as the breaking angle with respect to the beach. Also, it has been found that the obstructions across the beach such as piers, breakwaters, and points influence the circulation pattern and sedimentation areas. In such cases, the obstruction determines the position of one side of the circulation cell.

Field observations of the nearshore circulation system show many interesting features. For instance, it has been found that the distance between rip currents decreases as the intensity of the breaker increases, indicating that the longshore current has some limiting value above which it breaks seaward into rip currents. The spacing between rip currents ranges between 30 to 1000 m and the maximum value of the longshore velocity appears not to exceed 2–3 knots. Calculations of the volume transported seaward by the rip currents indicate that along straight shores, the discharge of water by those currents is between 2 and 10 percent of the total estimated volume transported shoreward by the waves between the rip currents.

Equations have been formulated that relate the longshore current velocity and the characteristics of the oncoming waves and of a beach drag coefficient. However, no satisfactory relationship has been developed between the rate of littoral transport and the waves and longshore currents that generate it. The amount of sediment transport along oceanic coasts is estimated from observed rates of erosion or deposition in areas where sediments can be monitored. The values obtained show that the transport rates vary from small values to a few million cubic meters per year. Laboratory experiments carried out by different investigators indicate that the values of littoral transport in models depend on wave energy, wave steepness, and the angle of wave approach. Also, with waves of low steepness ratio, the transport takes place in the wave-swash zone; while with steeper waves, the sediment transport occurs principally on the longshore bar in the breaker zone. Moreover, as the beach is always being adjusted to the waves that shape it, marked changes in transport rate were observed in the process of this adjustment.

In the present state of the art, the recommended approach to predict the sediment movements seems to be consideration of how the incoming energy is transferred and dissipated. The most promising line of attack for both field and laboratory investigation of nearshore processes has been the development of multichannel data-acquisition systems and the computer techniques necessary for the rapid treatment of time series of observations into simple terms of power, spectra, cospectra, and phase differences.

SUMMARY

The description of the sea bottom or benthic boundary is the main purpose of submarine geology. The relief of the ocean shows the presence of three major features: the continental margin, the ocean basin floor, and the midoceanic ridge. The continental margin is the province of greatest interest to the engineers because present technology is available for the exploitation of its resources and because a considerable percentage of the world population lives along the continental boundary and is making an ever-increasing use of it. Sediment properties and dynamics is the other subject of submarine geology of considerable scientific and engineering interest. Nearshore processes, which include the phenomena taking place at the interface between aerial landscape and sea present some of the most challenging problems to the engineer. The central problem here is the understanding of sediment motions, transport, and deposition. Any man-made construction may have very undesirable effects if not properly designed.

As was mentioned in Chapter 1, the responsibility of the engineer falls into two areas, namely (1) designing and developing the instruments and tools for environmental studies and (2) designing and developing the tools so that they may be used effectively.

With respect to the benthic boundary, the following tasks lie ahead for the engineer:

1. Environmental studies on or near the bottom: This includes information on currents, erosion, and mass flows on the sea floor to aid in design or prediction. Fine detail of bottom relief, such as ripples, small-scale topography, and objects are practically unknown except in a few particular places. It is necessary to develop and use techniques for continuous measurements, visual observations and collections and samplings on transects across the bottoms of the several seas, and it is needed to

analyze, describe, and typify these observations. The approach should be general and versatile. It should concentrate on flexible visual observation, generalized measurement, full-range sampling, and recovery ability. The present approach to conventional measurement and sampling from surface ships and photographs at discrete points is unsatisfactory because of the inability to follow and observe instantly some particular interesting area or object and because of its clumsy and limited ability to collect this information. It is unclear whether such explanation could best be conducted by a free manned search vehicle and mother ship or by an instrument or manned pod towed near the bottom by a surface or near-surface vessel. Since accurate navigational control is essential to all classes of surveys, worldwide navigation with an accuracy of 0.1 nautical mile is desirable.

2. Mechanical properties of sediments: This is a basic need for a rational design of construction on the sea floor. Most of the information available comes from submarine geological description and geochemical and biogeographical examination. Few have the precision of the terrestrial (subaerial) measurements. Because of the difficulties experienced in relieving the pressure on the sample brought to the surface, development of instruments and techniques for in situ determination of structural characteristics is needed. A partial list of specific needs in ocean bottom construction includes prediction of the following: support of bottom vehicles and installations; penetration of waste-disposal containers, covers, weaponry, nose cones, anchors, downed submarines, archeological artifacts, meteorites, manganese modules, and dredges; stability of slopes; erodibility around installations and their discharges; and holding ground for moorings and drill-hole reentry foundations.

3. Nearshore processes: These are the problems related to the uses and effects of the coastal environment. The uses are recreation; source of industrial and drinking water; dumping ground for garbage, sewage, industrial, and agricultural wastes; coastal engineering projects such as harbors; and connection between sea transportation and networks on land. The effects are erosion by ambiential factors, deposition, pollution of harbors and bays, oil-covered beaches, and so on.

Glossary

Absorption   The loss of sound or light intensity caused by the conversion of sound or light energy into heat as it passes through air or water.

Abyssal cone   Larger feature than a deep-sea fan formed by a family of overlapping deep-sea fans. They are found off the big rivers, such as the Mississippi and Congo.

Adiabatic   A process by which a change of state (i.e., of the temperature, pressure, and density) of a fluid takes place without heat being applied or withdrawn. It is also called the isentropic process. In meteorology and oceanography, the main interest centers in the changes of the vertical distribution of temperature by changes of pressure.

Advection   The process of the transfer of air or water by horizontal motion.

Afternoon effect   Shallow thermocline or negative thermal gradients produced by the daily heating which results in a downward refraction of sound rays and, consequently, a reduction of sonar ranges.

Amphidromic (point)   A point in the ocean of zero tidal range. The generation of this particular point, around which the crests and troughs of the edge waves (Kelvin wave) seems to rotate, is related to the propagation of the tidal wave affected by the rotation of the earth (Coriolis acceleration). This amphidromic point is believed to exist in many closed basins in which the period of standing co-oscillation of the tide and Kelvin wave components can remain 90 degrees out of phase.

Attenuation   The reduction in sound or light intensity caused by the absorption and scattering of sound or light energy in air or water.

Autotrophic   Phytoplankton that require no organic material whatever for their normal growth and reproduction.

Baroclinity   Condition where the isosteric and isobaric surfaces do not coincide.

Barotropy   Condition where the isosteric and isobaric surfaces coincide. Density does not vary along isobaric surfaces.

Benthic division   A primary biotic division of the sea which comprises all of the ocean floor. The benthos is constituted of the animals and plants which live on or in the bottom of the ocean and also those whose lives are in some way closely connected with the ocean bottom.

Benthos   The category of marine organisms that live on, in, or close to the bottom of the ocean.

Biota   Refers to all fauna and flora.

Caballing   Sinking of a water parcel that resulted from mixing of two water types of different salinity and temperature but of the same density.

Cnoidal Waves   Waves that are periodic in form and which tend to become the solitary wave (no change of form) in the limiting case of long wave lengths. They constitute a system of oscillatory waves of finite height propagating in a canal of limited depth.

Continental rise   A gentle slope with a generally smooth surface found at the base of the continental slope.

Decibar   Pressure exerted by a column of pure water of a height of 1 dynamic meter. It is a tenth part of a bar, which is defined as 10^6 dynes per sq cm.

Deep scattering layer (DSL)   A layer or layers in the ocean believed to consist of plankton and fishes from which sound rays are scattered.

Deep-sea fan   A sedimentary feature formed by the sediments carried by turbidity currents and successively deposited at the mouth of a single deep-sea canyon.

Density (potential)   Density which would occur in a water mass after adiabatic displacement of the mass with its in situ temperature and salinity from the depth to the surface.

Dynamic meter   The work required to lift a unit mass nearly 1 linear meter against gravity. It is a dynamic unit equal to 10^5 dyne-cm per g.

Eddy coefficients   Exchange coefficients (of momentum, heat, and salt) determined for turbulent flow.

Edgewave Widely spaced wave crests arranged at right angles to the shoreline. Such waves may be excited by a wind shift associated with a passing front.

Euphotic zone Surface layer of the ocean in which photosynthesis takes place.

Eustatism The fluctuations of sea level due to the changing capacity of the ocean basins or the volume of ocean water. The agents producing these changes may be glaciation or melting of glaciers, sedimentation filling the ocean basins, and tectonics affecting the size of the ocean basins.

Extinction coefficient (absorption coefficient) Coefficient that measures the rate of decrease of downward-traveling radiation in sea water.

Froude number The ratio of the inertial term to the gravitational term.

Geopotential surface See Level surface.

Geostrophic motion Fluid flow that is both unaccelerated and frictionless. In this case, the pressure gradient force is balanced by the Coriolis force and gravity. The technique used to derive the geostrophic motion is known as the *geostrophic approximation.*

Globigerina A very small marine animal of the foraminifera order, with a chambered shell; or the shell of such an animal. In large areas of the ocean, the calcareous shells of these animals are very numerous, being the principal constituent of a soft mud or globigerina ooze forming the ocean bed.

Halocline A steep vertical gradient of salinity.

Ideal sea level The most important gravitational level surface; also known as geoide.

Inertia stability The property by which a parcel of a fluid in geostrophic motion along straight contours resists being moved across contours northward or southward.

Inertia waves Waves produced when large-scale perturbations are considered which makes the Coriolis force important.

Isentropic surface A surface of constant potential temperature.

Isobaric surface A surface of constant pressure.

Isobath Depth contour.

Isopycnic surface Surfaces of constant density.

Isosteric surfaces Surfaces of constant specific volume.

Laminar flow A flow in which the fluid motion is highly organized and moves smoothly in streamlines in parallel layers or sheets; a nonturbulent flow.

Level surface A surface along which no component of gravity acts and is, therefore, perpendicular to the force of gravity.

Molecular coefficients Exchange coefficients (of momentum, heat, or salt) determined in a laboratory for laminar flow.

Nutrient (salts) Mineral salts dissolved in the ocean that have been shown to influence and, at times, limit or control the production of phytoplankton, namely, the nitrogen and phosphorus and, also, to a more uncertain degree, iron and other trace elements.

Orthogonals (ray) Lines drawn perpendicular to the wave crests. Very commonly used in wave refraction diagrams.

Pendulum day Time required for a complete turning of the plane in which a Foucault pendulum oscillates. It is equal to 24 hr divided by the sine of latitude.

Planetary vorticity Vorticity imparted to a resting fluid—relative to an inertial system—by the component of the earth's rotation around the local vertical. The Coriolis parameter, f, is a measure of the planetary vorticity at any latitude.

Polymerization The characteristic of certain chemical compounds to have some individual molecules clustered together. This behavior affects some physical properties.

Potential density The density of sea water for the in situ salinity and potential temperature. In other words, it is the density of a parcel of water of certain salinity and at certain depth when brought adiabatically to the sea surface.

Potential temperature The temperature that a parcel of water would attain if raised adiabatically from the depths to the sea surface.

Power spectrum The description of a train of waves showing the distribution of energy among waves of different periods. In oceanography, this is also known as *wave spectrum*.

Pycnocline A steep vertical gradient of density.

Reynolds number The ratio of the inertial term to the frictional term. In any given motion it indicates whether inertia or viscous effects predominate.

"Roaring Forties"   The area of the oceans between 40° and 50° south latitude where strong westerly winds prevail.

Rossby number   The ratio of convective accelerations to the Coriolis force which provides an overall estimate of the relative importance of the nonlinear terms. It also provides an indication of the degree of interaction between the steady-state and the time-dependent terms.

Scattering   The random dispersal of sound or light energy produced by the reflection of the sound and light rays by small particles, irregularities in the reflecting layers, or other scatterers.

Seiche   A stationary wave oscillation with a period varying from a few minutes to an hour or more but somewhat less than the tidal periods. They are usually attributed to strong winds or changes in barometric pressure and are found both in enclosed bodies of water and superimposed upon the tide waves of the open ocean.

Short-crested waves   An ocean wave whose crest is of finite length, i.e., the type actually found in nature.

Sound channel   That region in the water column where the sound velocity first decreases to a minimum value with depth and then increases in value.

Stratosphere (oceanic)   The lower part of the thermal stratification in the ocean and extending from the bottom of the troposphere (thermocline) down to the sea bottom and is characterized by very small changes in temperature both in horizontal and vertical directions.

Substantial derivative (individual time change—Eulerian derivative)   Measure of the rate of change of one particular property of one parcel of fluid with time. It implies the knowledge of the three-dimensional trajectory of the parcel and then the measurement of the change of the property along the trajectory.

Surf   The wave activity in the area between the shoreline and the outermost limit of breakers.

Troposphere (oceanic)   Upper layers of the oceans extending from the surface down to about 600–1000 m. These layers are characterized by strong temperature decrease with depth.

Tsunamis   A long-period sea wave produced by a submarine

earthquake or volcanic eruption, commonly misnamed a tidal wave.

Turbulent flow   Fluid motion in which random motions of parts of the fluid are superimposed on a simple pattern of flow. The opposite is *laminar flow* or *streamline flow.*

Väisälä–Brunt frequency (stability freuqency)   Natural frequency of oscillation—adiabatic—of a vertical column of fluid given a small displacement from its equilibrium position. Its distribution provides a measure of the vertical stability of the fluid and is one of the most important characteristics of the ocean.

Water mass   Body of water that is represented by a definite T–S relationship.

Water type   Body of water which is represented on a T–S diagram by a single point.

Zonal (flow)   Flow along the parallel latitude—east-west direction or opposite.

Selected Readings

CHAPTER 2

A Draft of a General Scientific Framework for World Ocean Study, Prepared for The Intergovernmental Commission by SCOR, 1964.

Brahtz, J. F., *Ocean Engineering,* Chapters 1, 2, and 6, New York: Wiley, 1968.

Defant, A., *Physical Oceanography,* Vol. I, Chapter 1, New York: Pergamon Press, 1961.

Effective Use of the Sea, Report of the Panel on Oceanography of the President's Science Advisory Committee, Washington, D.C., 1966.

National Academy of Sciences Publication 1492—Oceanography 1966, Washington, D.C., 1967.

Neumann, G., and W. Pierson, Jr., *Principles of Physical Oceanography,* Chapters 1, 2, and 5, Englewood Cliffs, N.J.: Prentice-Hall, 1966.

Sverdrup, H. U., M. W. Johnson, and R. H. Fleming, *The Oceans*, Chapters 1, 2, 7, 16, and 17, Englewood Cliffs, N.J.: Prentice-Hall, 1946.

CHAPTER 3

Brahtz, J. F., *Ocean Engineering,* Chapters 6 and 7, New York: Wiley, 1968.

Defant, A., *Physical Oceanography,* Vol. I, Chapters 1–6, New York: Pergamon Press, 1961.

Dietrich, G., and K. Kalle, *General Oceanography,* Chapters 2, 4, and 5, New York: Wiley-Interscience, 1963.

Eckart, C., *Hydrodynamics of Oceans and Atmospheres,* Chapters 1, 2, 4, and 5, New York: Pergamon Press, 1960.

Hill, M. N., *The Sea,* Vol. I, Chapters 1, 2, 8, 9, 11–14, New York: Wiley-Interscience, 1962.

McLellan, Hugh J., *Elements of Physical Oceanography,* Chapters 4–6, 16–18, New York: Pergamon Press, 1965.

Neumann, G., and W. J. Pierson, Jr., *Principles of Physical Oceanography,* Chapters 3, 9, 13, and 14, Englewood Cliffs, N.J.: Prentice-Hall, 1966.

Sverdrup, H. U., M. W. Johnson, and R. H. Fleming, *The Oceans,* Chapters 3–5 and 15, Englewood Cliffs, N.J.: Prentice-Hall, 1946.

Von Arx, W. S., *Introduction to Physical Oceanography,* Chapters 1, 5–7, Reading, Mass.: Addison-Wesley, 1962.

CHAPTER 4

Defant, A., *Physical Oceanography,* Part II, Chapters 9–21, New York: Pergamon Press, 1961.

Dietrich, G., and K. Kalle, *General Oceanography,* Chapters 7 and 10, New York: Wiley-Interscience, 1957.

Eckart, C., *Hydrodynamics of Oceans and Atmospheres,* New York: Pergamon Press, 1960.

Fofonoff, N. P., *The Sea,* Vol. I, Sect. III, New York: Wiley-Interscience, 1962.

Greenspan, H. P., *The Theory of Rotating Fluids,* Cambridge, Mass. Harvard University Press, 1968.

Hendershott, M. C., *Ocean Engineering,* Chapter 7, New York: Wiley, 1968.

McLellan, H. J., *Elements of Physical Oceanography,* New York: Pergamon Press, 1965.

Neumann, G., and W. J. Pierson, Jr., *Principles of Physical Oceanography,* Chapters 6–8, Englewood Cliffs, N.J.: Prentice-Hall, 1966.

Panofsky, H., *Introduction to Dynamic Meteorology,* Chapters 2 and 4, University Park: Pennsylvania State University, 1958.

Phillips, N. A., "Geostrophic Motion," *Rev. Geophys.,* Vol. 1, No. 3, pp. 123–176, 1963.

Phillips, O. M., *The Dynamics of the Upper Ocean,* Chapters

1 and 2, Cambridge, Mass.: Harvard University Press, 1966.

Robinson, A. R., *Wind-Driven Ocean Circulation*, Waltham, Mass.: Blaisdell, 1963.

Stommel, H., *The Gulf Stream*, Berkeley: University of California Press, 1965.

Sverdrup, H. U., M. W. Johnson, and R. H. Fleming, *The Oceans*, Chapters 11–13, Englewood Cliffs, N.J.: Prentice-Hall, 1942.

Von Arx, W. S., *Introduction to Physical Oceanography*, Chapters 4, 6, and 7, Reading, Mass.: Addison-Wesley, 1962.

CHAPTER 5

American Society of Civil Engineers Coastal Engineering Research Council, *Proceedings of Conferences on Coastal Engineering*, No. 1–12, 1950–1966.

American Society of Civil Engineers, *Proceedings of the Conference on Civil Engineering in the Oceans*, San Francisco: September 6–8, 1967.

Brahtz, J. F., *Ocean Engineering*, Chapter 7, New York: Wiley, 1968.

Defant, A., *Physical Oceanography*, Vol. II, New York: Pergamon Press, 1961.

Hill, M. N., *The Sea*, Vol. I, Chapters 15–23, New York: Wiley, 1962.

Kinsman, B., *Wind Waves*, Englewood Cliffs, N.J.: Prentice-Hall, 1965.

National Academy of Sciences, *Proceedings of a Conference in Ocean Wave Spectra at Easton Maryland, May 1–4, 1961*, Englewood Cliffs, N.J.: Prentice-Hall, 1963.

Phillips, O. M., *The Dynamics of the Upper Ocean*, Chapters 3–5, Cambridge: Harvard University Press, 1966.

Stoker, J. J., *Water Waves—The Mathematical Theory with Applications*, New York: Wiley-Interscience, 1957.

Von Arx, W. S., *Introduction to Physical Oceanography*, Chapter 3, Reading, Mass.: Addison-Wesley, 1962.

Wiegel, Robert L., *Oceanographical Engineering*, Chapters 1–12, Englewood Cliffs, N.J.: Prentice-Hall, 1964.

CHAPTER 6

Brahtz, J. F., *Ocean Engineering*, Chapter 8, New York: Wiley, 1968.

Forchhammer, G. V., On the Composition of Sea Water in the Different Parts of the Ocean, *Phil. Trans. Roy. Soc.* Vol. 155, 203–262, 1865.

Harvey, H. W., *The Chemistry and Fertility of Sea Waters*, Cambridge, Mass.: Harvard University Press, 1963.

Hill, M. N., *The Sea*, Vol. 2, Chapters 1–9, 17–21, New York: Wiley-Interscience, 1963.

Mero, J. L., *The Mineral Resources of the Sea*, New York: American Elsevier, 1965.

Raymont, J. E. G., *Plankton and Productivity in the Oceans*, New York: Pergamon Press, 1963.

Riley, J. P., and G. Skirrow, *Chemical Oceanography*, New York: Academic Press, 1965.

Tresler, D. K., and J. W. McLemon, *Marine Products of Commerce*, New York: Reinhold, 1951.

CHAPTER 7

American Society of Civil Engineers, Coastal Engineering Research Council, *Proceedings of Conferences on Coastal Engineering*, Nos. 1–11, 1950–1968.

Guilcher, A., *Coastal and Submarine Morphology*, Great Britain: Butler & Tanner, Ltd., 1958.

Hill, M. N., *The Sea*, Vol. 3, Chapters 5, 12–14, 16, 17, 20, 21, 27, and 29, New York: Wiley-Interscience, 1963.

Ippen, A. T., *Estuary and Coastline Hydrodynamics*, Chapter 9, New York: McGraw-Hill, 1966.

Johnson, J. W., "Dynamics of Nearshore Sediment Movement," *Bull. Am. Assoc. Petrol. Geol.*, Vol. 40, No. 9, September 1956.

King, C. A. M., *Beaches and Coasts*, London: E. Arnold, 1959.

Shepard, F. P., *Submarine Geology,* New York: Harper & Row, 1963.

Von Arx, W. S., *Introduction to Physical Oceanography*, Chapter 2, Reading, Mass. Addison-Wesley, 1962.

Wiegel, R. L., *Oceanographical Engineering*, Chapter 14, Englewood Cliffs, N.J.: Prentice-Hall, 1964.

ADDITIONAL READING

Myers, John J., C. H. Holm, and R. F. McAllister, *Handbook of Ocean and Underwater Engineering*, New York: McGraw Hill, 1969.

Bretschneider, Charles L., Editor, *Topics in Ocean Engineering*, Houston, Texas: Gulf Publishing Co., 1969.

Index